fundamentals of odor perception • fragrance and psychophysiology • olfactory conditioning • scent and social behavior • mood • cultural and historical perspectives • applications • fundamentals of odor perception • fragrance and psychophysiology • olfactory conditioning • scent and social behavior • mood • cultural and historical perspectives • applications • fundamentals of odor perception • frag fundamentals of odor perception • fragrance and psychophysiology • olfactory conditioning • scent and social behavior • mood • cultural and historical perspectives • fundamentals of odor perception • fragrance and psychophysiology • olfactory conditioning • scent and social behavior • mood • cultural and historical perspectives • applications • fundamentals of odor perception • fra

Compendium of Olfactory Research

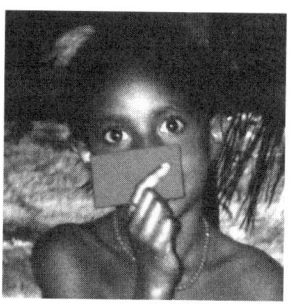

Explorations in Aroma-Chology: Investigating
the Sense of Smell and Human Response to Odors
1982-1994

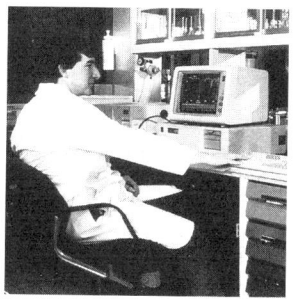

Edited By
Avery N. Gilbert

Olfactory Research Fund, Ltd.

KENDALL/HUNT PUBLISHING COMPANY
4050 Westmark Drive Dubuque, Iowa 52002

Foreword

Dr. Jack Mausner
Chairman
Olfactory Research Fund, Ltd.

This "Compendium of Aroma-Chology Research" represents the dawn of a new era in olfactory research. The scientific investigations of the last two decades of the 20th century explored the positive effects of fragrance on human behavior and have opened new vistas of olfactory research for scientists to explore and conquer in the 21st century. Aroma-Chology has proven to be more than a New Age fad that will fade – not only has it become an integral part of the global mainstream, but both industry and individuals are routinely incorporating this new science in concepts for the future. Architects, furniture designers and interior decorators use it as an intrinsic element in their designs. We now see airports, hotels, schools, factories, corporate offices, hospitals and nursing homes utilizing fragrances for the purpose of creating environments that are conducive to relaxation, learning, increased performance and alertness. In addition, individuals are using scents as a way of empowering themselves to take control of their own feelings of well-being by managing their moods and emotions.

The research studies supported by the Olfactory Research Fund to date, and reported in this volume, indicate that we must look at fragrance for more than its purely hedonic value

for there is clear evidence that it possesses a much more functional value and can have a very positive effect on many aspects of our daily lives. We now know, without a doubt, that aromas have a direct effect on mood and hence, mood-related behavior. Likewise, research findings have shown that fragrances can successfully reduce stress and anxiety, increase alertness and performance, and have a beneficial effect on sleep patterns. Investigations have also demonstrated the significant role fragrances play in our social relationships – between members of the opposite sex, mother and child, siblings, co-workers and the myriad other social relationships we encounter daily.

As science further validates the link between the sense of smell and the many other components which contribute to our overall well-being, consumers will increasingly demand products with multi-sensory benefits offering the potential to improve their quality of life. The trailblazing work reported here has put us well on our way to meet that demand.

The Olfactory Research Fund will continue to significantly support further investigations into the hidden powers of the sense of smell and disseminate the results of this research. Indeed, as interest in the science of smell continues to expand internationally, the next "Compendium of Aroma-Chology Research" will certainly include reports of research conducted in major hospitals and universities, not only in North America but around the world. It is predicted that olfaction may well be the universal language that unites all of us in the global village of the 21st century.

Historical Perspective

Annette Green
President
Olfactory Research Fund, Ltd.

As we stand on the brink of the 21st Century, we find ourselves in the midst of a sensory revolution. Driven by scientific research, it has provided a range of ground-breaking information about our least understood sense: **smell.**

Although scientists have sought to unlock the mysteries of the sense of smell for centuries, it was not until the latter part of the 20th century that isolated academic environments focused their research efforts on the chemical senses — taste and smell. In 1981, an important step forward was taken. The Board of Directors of The Fragrance Foundation, a non-profit, educational arm of the fragrance industry, voted to create an independent, tax-exempt charitable organization, with the unique mission of exploring the sense of smell and the effects of odor on human behavior. The first Board of Directors was appointed and Richard B. Salomon was elected President.

The mandate of the Olfactory Research Fund was, and still is, as stated in its By-Laws, to sponsor significant scientific research and promote the study of the olfactory arts and sciences; sponsor and conduct educational and public outreach programs which promote and stimulate the study and understanding of the sense of smell. Today, the Fund is internationally recognized for its innovative role in supporting and advancing sensory research.

As the Fund evolved, the focus of the research it supports has shifted from basic research to the science of Aroma-Chology, a term which was coined by the Olfactory Research Fund to describe the study of the inter-relationship of psychology and the latest in fragrance technology to explore the positive effects of fragrance on human behavior.

To date, the Olfactory Research Fund has provided grants to over twenty-five clinical psychologists, researchers and anthropologists. The results of their research reveal that fragrance can positively affect sleep, stress, alertness, social interaction and general feelings of well-being. This journal contains reports on each of these studies.

The Fund continues to explore new olfactory frontiers and, each year, solicits grant proposals for innovative research projects which seek to integrate the study of olfaction with current issues in developmental, social and cognitive psychology, as well as the related disciplines of anthropology and sociology.

The following time line highlights the important milestones which mark the growth of the Olfactory Research Fund and the establishment of Aroma-Chology as a significant branch of olfactory research as well as the assimilation of Aroma-Chology into the mainstream of society.

Olfactory Research Fund Time Line

1980

January 31st – Annette Green, Executive Director of The Fragrance Foundation proposed the idea of forming a tax-exempt charitable organization dedicated to funding research on the sense of smell.

1981

January 26th – Certificate of Incorporation of the Fragrance Foundation Philanthropic Fund approved by state law.

September 16th – First board meeting of the newly created Fragrance Foundation Philanthropic Fund attended by directors: Richard B. Salomon, William S. Cain, Fernando Aleu, George Friedman, Eugene Grisanti, Alvin Lindsay, Eugene Milano, Bernard Mitchell, Richard Lockman and Annette Green.

September 28th – The Fragrance Foundation Philanthropic Fund, Ltd. registered as a Charitable Corporation 501(c)3.

Richard Salomon appointed first president of the newly created Fragrance Foundation Philanthropic Fund.

First contribution made to the Fund in the amount of $5,000 by Richard Salomon.

Additional contributions in the amounts of $5,000 each made by Jovan and The Fragrance Foundation.

1982

May 26th – First symposium on the sense of smell held at the Waldorf-Astoria, New York.

1983

August 10th – Board of Directors awarded the first two grants to: Dr. Susan Schiffman, Duke University and Dr. Howard Ehrlichman, CUNY Research Foundation, City University of New York.

Richard Salomon made personal contribution of $250,000 to Monell Chemical Senses Center in the Fragrance Foundation Philanthropic Fund's name.

The Fragrance Foundation Philanthropic Fund granted $100,000 to Monell Chemical Senses Center.

1984

March – Dr. Fernando Aleu elected second President of the Fund.

June 26th – Sense of Smell Award designed by Pierre Dinand is approved by Fragrance Foundation Philanthropic Fund Board.

October 1st – First "Sense of Smell" Awards luncheon held at the Waldorf-Astoria Hotel.

October 1st – First Scientific Sense of Smell Award given to Dr. Lewis Thomas, President, Emeritus, Memorial Sloan-Kettering Cancer Center.

October 1st – First Retail Sense of Smell Award given to Marvin Traub, Chairman, Bloomingdale's.

November 1st – Fragrance Foundation Philanthropic Fund Board of Directors voted to change name of the organization to "Fragrance Research Fund, Ltd."

1985

June 4th – The first "Medals of Honor" contributors were recognized at the Thirteenth Annual Fragrance Foundation Recognition Awards.

1986

July – Annette Green attended the First International Perfumery Congress held in Warwick, England as a keynote speaker representing the Fragrance Research Fund.

Industry and Scientific Advisory Committees formed.

Annette Green coined the term "Aroma-Chology" to identify new science that explores the inter-relationship of psychology and the latest in fragrance technology to transmit through odor a variety of positve feelings (e.g., relaxation, exhilaration, sensuality, happiness & achievement).

1987

September – First issue of the Fragrance Research Fund Newsletter, "Focus on Fragrance," is published.

1988

November 16th – The Board of Directors of the Fragrance Research Fund elects Dr. Jack Mausner, Sr. Vice President, Research and Development, Chanel Inc., as its new President.

1989

July – Registration of a service mark, "Aroma-Chology," is approved by U.S. Patent & Trademark Office.

1991

November – First Aroma-Chology Symposium, which featured a wide range of international experts in the field of Aroma-Chology, sponsored by the Fragrance Research Fund held at the Macklowe Theatre, New York.

1992

January – Fragrance Research Fund renames newsletter, "Focus on Fragrance," to the "Aroma-Chology Review."

October 20th – Registration of a service mark, "Aroma-Chology Information Center," is approved by U.S. Patent & Trademark Office.

October 30th – Fragrance Research Fund Board of Directors voted to change name of the organization to "Olfactory Research Fund, Ltd."

1993

January – Olfactory Research Fund established a new program to raise funds for public educational outreach programs called "The Fifth Sense Commendation."

April 12th – Certificate of Incorporation amended to reflect name change of the Fragrance Research Fund to the Olfactory Research Fund.

June – Olfactory Research Fund received first pledge to the "Fifth Sense Commendation" program. First $100,000 contribution ever received from a corporation (Quest International Inc.).

September - Olfactory Research Fund "Get in Touch With Your Sense of Smell" public service ad campaign launched in the New York Times.

1994

May 10th – Registration of a service mark "Get in Touch with Your Sense of Smell," is approved by U.S. Patent & Trademark Office.

June 11th – First Annual National Sense of Smell Day sponsored by the Olfactory Research Fund is celebrated in Science Centers/Museums in 13 cities.

November 1st – First Annual "Night of Honors" celebrating the sense of smell held at the United Nations, New York.

November 1st – First Richard B. Salomon Award presented to Diane Ackerman, Best-Selling Author of "A Natural History of the Senses" and to WETA, public service television for producing mini-series based on Ms. Ackerman's book.

Editor's Note

Avery N. Gilbert
Scientific Affairs Director
Olfactory Research Fund, Ltd.

The sense of smell remains the great uncharted frontier of sensory science. Certain major features of its natural history are known, but others no doubt await discovery. Even after they have been found, there will be many challenges to meet before we can claim a comprehensive biological and psychological theory of olfaction.

Perfumers and perfumery chemists were among the first to empirically explore this realm: their trade existed for centuries before the invention of experimental sensory psychophysics in the nineteenth century. Much of what we know today can be traced to the pioneering, but not strictly scientific, investigations of perfumers.

Having done research in an academic setting as well as in the world of commercial fragrance, I have a great appreciation for the know-how — the accumulated sensory knowledge — of the industry's professionals. Their insights can serve as the spark for truly creative sensory research, and they in turn can find their work enhanced by its results.

In this spirit of art leading science, the Olfactory Research Fund has alerted academic scientists to the great potential for discovery that lies in olfaction. Researchers have responded, and studies sponsored by the Fund, many described in these pages, have helped fill the gaps in our knowledge.

The reports in this book have been organized into sections according to seven themes. I have provided an introduction to each section, in order to place the reports into a larger context for the reader.

Avery N. Gilbert

Dr. Avery Gilbert is the Scientific Affairs Director of the Olfactory Research Fund and a member of the Scientific Affairs Committee of the Fragrance Materials Association. Dr. Gilbert describes himself as an evolutionary biopsychologist. He investigates how social and psychological factors interact with odor perception, including how expectation can bias odor judgments. He is also interested in how Alzheimer's disease and normal aging alter the sense of smell. His most recent work explores cross-modal associations of fragrance to color and sound. Dr. Gilbert co-authored the National Geographic's Smell Survey – the largest participatory experiment of its kind. Dr. Gilbert's work is frequently cited in newspapers and magazines. Dr. Gilbert is the founder and President of Synesthetics, Inc., a firm specializing in sensory research and communication.

Acknowledgements

The Compendium of Olfactory Research is the first of its kind to be published. It represents the results of over ten years of research supported by the Olfactory Research Fund. Sincere appreciation must begin with the corporations and individuals whose contributions and encouragement, since the Fund was established in 1981, have enabled us to sponsor the research reported here.

The Fund especially wishes to thank and praise the scientists who have forged frontiers in olfaction and laid the foundation for the new science of Aroma-Chology. Their reports summarizing this research comprise this definitive body of work.

Sincere thanks to Avery Nelson Gilbert, Ph.D., who has generously contributed his expertise as editor of the Compendium. His assistance in organizing the wealth of material contained in these pages has been invaluable.

Graphic Design (cover and text): Julia Ptasznik Visual Communications.

Table of Contents

IV. Scent and Social Behavior

How scent affects interpersonal attitudes and behavior.

V. Mood

How odor impacts mood.

VI. Cultural and Historical Perspectives

*The different ways scents are used across cultures
and times.*

VII. Applications

How research results can be put to work.

Olfactory Research Fund, Ltd.

Compendium of
Aroma-Chology Research
1982-1994

Fundamentals of Odor Perception

physiology • olfactory co

nd social behavior • mood • cultural and historical perspectives • applications • fundamentals of odor per

and psyc

I

nentals of

perception • fragrance and psych

The initial event in odor perception takes place in the nose, where the nervous system transforms chemical information (volatile fragrance molecules) into electrical information (nerve impulses). This information is sent to the brain via the olfactory bulb, a complex anatomical structure that is the first processing point for odor information. Charles Greer has studied the olfactory bulb, and his findings will help scientists understand its cellular structure and circuitry, and possibly account for sensory and psychological features of smell.

As we smell our food and sample perfumes, the process of odor perception feels seamless and simple: smells just happen. But how does the swirl of fragrant molecules in our nose become a stable pattern of odor perception? David Laing studied the perception of odor mixtures, and the phenomena of suppression and blending. His results indicate that odors arriving as little as a tenth of a second apart alter the perception.

Are we born with built-in preferences for certain smells, or are they acquired through experience? New answers to this fundamental question come from Hilary Schmidt's work on the development of odor perception in infancy and early childhood. She invented new ways to measure the olfactory likes and dislikes of children as young as six to nine months, and discovered that their preferences were similar to those of adults.

People differ widely in their sensitivity and preference for smells. The potential causes of this variation are not well known. They include learning and experience, as well as biological differences, some of which may be genetic. Nancy Segal compared the odor perception of identical and fraternal twins to estimate how much variation can be attributed to genetic similarity.

Anatomical Organization of the Human Olfactory System

Charles A. Greer, Ph.D.
Yale University School of Medicine

Our studies are providing a much clearer understanding of the cellular and circuit organization of the human olfactory bulb. An important notion that has been reinforced by these studies is the inherent complexity of the human olfactory system. The colloquial suggestion is often made that the complexity, sensitivity and efficiency of the human sense of smell is less than other species because of our "atrophied olfactory organs." The research summarized below does not support such a suggestion. Indeed, our studies have clearly shown a level of complexity in organization and heterogeneous composition of synaptic circuits that is equivalent to other species.

A particular important development that has grown out of our work is the hypothesis that olfactory deficits associated with specific disease processes may have well defined olfactory bulb pathologies. Most clear in this regard is the possible relationship between olfactory deficits in Parkinson's Disease and the population of TH immunoreactive (dopaminergic) juxtaglomerular neurons. Our research demonstrating a population of TH-like immunoreactive neurons in the human olfactory bulb is the first to recognize that these neurons may represent a specific dopaminergic mechanism, in common with the basal ganglia, and able to account for the smell disorders found in patients with Parkinson's Disease.

In a parallel vein, it may well be that utilization of fundamental neuropharmacology targeted for specific olfactory bulb neurons/circuits may be effective in increasing or decreasing olfactory acuity. For example, the dopaminergic antagonist haloperidol may be effective in decreasing olfactory acuity in a fashion that parallels Parkinson's. In contrast however, dopaminergic agonists such as apomorphine may then be a potential neuropharmacological strategy for increasing olfactory acuity. These and related hypotheses regarding the relationship between the fundamental organization of the human olfactory system and function are important questions that we look forward to being able to address in the future.

During the 12 months of our funding from the Olfactory Research Fund, we were successful in obtaining human olfactory bulbs for analyses. Three primary sources were utilized: 1) nonpathological surgical specimens obtained during the course of neurosurgical protocols that necessitated unilateral removal/sacrifice of the olfactory bulb; 2) non-pathological specimens obtained during autopsies performed at Yale University School of Medicine; and 3) specimens generously provided by the Los Angeles National Neurological Research Specimens Bank. Using these tissues we have completed our studies of the distribution and characteristics of neuronal somata and processes exhibiting immunoreactivity for olfactory marker protein (OMP), tyrosine hydroxylase (TH), somatostatin (SS), substance P (SP), serotonin (5HT), cholecystokinin (CCK), calcium binding protein (CaBP), parvalbumin (PV) and synaptophysin (SYN). The results for each of these will only be summarized below, full details are available in 1 or more of the manuscripts/publications included as part of the appendix.

Laminar Organization of the human olfactory bulbs obtained at postmortem or during surgical procedures was comparable. Cresyl violet staining demonstrated a distinctive laminar organization that paralleled prior descriptions in other species. Five layers were unequivocally identified: 1) the olfactory nerve layer that consists primarily of the axons of olfactory receptor cells; 2) the glomerular layer, a series of demarcated spherical regions of neuropil where the primary synapses between receptor cell axons and target dendrites occurs; 3) the external plexiform layer containing both the cell bodies of primary projection neurons and interdigitated zones of neuropil where local circuit interactions occur between the projection neurons and interneurons; 4) the mitral cell layer, a transition zone marking the deep boundary of the external plexiform layer and often containing the cell bodies of the primary projection neuron, mitral cells; and 5) the granule cell layer containing the somata of the largest population of interneurons, granule cells, and the interspersed exiting axons of projection neurons as well as arriving centrifugal axons. Finally, unique to the human and other primates, the anterior olfactory nucleus is found

deep to the granule cell layer in the posterior 2/3 of the olfactory bulb.

Olfactory Marker Protein localized to the glomeruli of the human olfactory bulb much as it does in other mammalian species. Staining with OMP revealed that the glomeruli of the human bulb are more widely dispersed than in the more commonly studied rodents. Of particular interest was the recognition that the fasciculation of individual axons travelling to a specific glomerulus occurred at some distance from the glomerulus, within the olfactory nerve layer. This seemingly simple observation provides some of the first evidence that the targeting of subsets of axons to a specific glomerulus may occur independent of the glomerulus; one hypothesis is that membrane-bound molecular markers on the axons may provide for mutual attraction or adhesion. This implies further, that the functional specificity of the glomerulus, as has been discussed from electrophysiological, 2-deoxyglucose and recently c-FOS studies of the rodent, may be determined by the axon fascicle and not necessarily by a pre-established glomerular map.

Synaptophysin-like immunoreactivity appeared somewhat similar to that of OMP in that the darkest staining was present within the glomeruli. SYN localizes to the 38,000 dalton protein characteristic of synaptic vesicles and the highest concentration or density of vesicles in the glomeruli. Lower levels of staining were also found in the external plexiform layer where SYN is believed to bind to the vesicle antigenic site in granule cell dendritic spines and mitral/tufted cell dendrites. Lower levels of staining were also seen in the granule cell layer, presumably reflecting the terminations of centrifugal axons.

Tyrosine Hydroxylase localized to a small subset of neuronal somata and processes in the glomerular layer. Based on the cross-sectional diameter of the somata and their restricted distribution surrounding glomeruli, the cells most likely correspond to the juxtaglomerular (periglomerular) neurons previously described in other species. What appeared to be dendritic processes, based upon their tapering profiles, were extensively distributed

within the glomerular neuropil. In contrast, the soma and axon-like processes were limited to the juxtaglomerular domain. As has been suggested in the rodent, there was no evidence of the immunoreactive axons leaving the glomerular layer, thus suggesting they may also be broadly classified as short-axon-neurons or local circuit neurons. Relative to other species, TH-like staining was somewhat less. Although all glomeruli appeared to have a circumferentially distributed population of immunoreactive processes, they were fewer in number than is typically seen in rodents. It is of great interest to speculate on the possible role of this particular population of neurons in diseases such as Parkinson's where the pathology of the basal ganglia is known to be limited to TH-containing dopamine neurons. Perhaps the olfactory deficits associated with the onset and progression of Parkinson's reflects a parallel loss of the TH-like immunoreactive neurons from the olfactory bulb.

Serotonin immunoreactivity was most apparent in the glomerular layer where stained varicose fibers appeared to integrate within individual glomeruli. Individual glomeruli varied extensively in the density of stained fibers; glomeruli with a dense plexus of 5HT immunoreactive fibers where the border of the glomerulus is highly defined by the stained fibers, intermingled with glomeruli that had only a few were stained fibers. A generalizable pattern of glomerular innervation was not apparent. A consistent feature of the glomerular 5HT immunoreactive fibers was their varicose or beaded appearance. In the deeper layers of the olfactory bulb intermittent varicose fibers were seen. Prior to reaching the glomerular layer these processes were not seen to bifurcate and appeared to follow relatively straight paths perpendicular to the olfactory bulb laminae. The density or distribution of varicosities along the length of the 5HT immunoreactive processes appeared comparable in the glomerular layer and remaining 3 deeper olfactory bulb layers. Somata immunoreactive for 5HT were not seen in the olfactory bulb. It is interesting to speculate that the variable density of 5HT fibers associated with any given

glomerulus may be a reflection of the odor specificity of that glomerulus. In this regard 1 testable hypothesis may be that the centrifugal modulation of odor processing by 5HT fibers may be variable for different odors.

Cholecystokinin immunoreactivity was similar, in many respects, to that described for 5HT. Deep in the subependymal and granule cell layers of the olfactory bulb CCK immunoreactive varicose processes were present. The processes were parallel to the olfactory bulb laminae in the subependymal zone but assumed a perpendicular orientation in the granule cell layer. Immunoreactive processes were also seen in the glomerular and external plexiform layers. In these more superficial layers the CCK positive fibers exhibited more diverse orientations, some of which were perpendicular to the laminae while others followed complex tortuous paths that frequently changed direction. In the glomerular layer the immunoreactive processes appeared to be limited to the juxtaglomerular region. No examples of CCK immunoreactive processes entering the intraglomerular neuropil were found. Somata immunoreactive for CCK were not evident in the olfactory bulb. Given the widely documented role of CCK in modulating appetitive behavior, including both eating and hunger, it is perhaps not surprising to recognize its potential importance in the modulation of odor processing. In a different vein, schizophrenics and the aged are both recognized to have diminished levels of CCK. Perhaps the alterations in satiety that accompany these conditions reflects, in part, an aberrant modulation of odors due to a decrease in CCK centrifugal modulation of olfactory bulb circuits.

Substance P immunoreactive processes were noted throughout the laminae of the olfactory bulb. Fiber-like processes had a varicose appearance and followed complex pathways in each of the laminae. In the subependymal core the fibers appeared parallel to the olfactory bulb surface, but in the granule cell layer a more perpendicular orientation was to delineate individual glomeruli by remaining restricted to the juxtaglomerular neuropil; there was no evidence of SP immunoreactive fibers entering into individual glomeruli. A few immunoreactive somata were found in the anterior olfactory nucleus. The somata ranged from 9-15um in diameter (mean = 11.6 + 0.79um). They appeared to be multipolar neurons based upon staining of the proximal processes. It was clear that the stained somata constituted only a small subpopulation of neurons in the anterior olfactory nucleus.

Somatostatin immunoreactivity was very similar to that of SP. Varicose fibers were found in all of the olfactory bulb laminae although in very low numbers. Individual fibers could be followed along the borders of glomeruli. There were no examples of SS immunoreactive fibers entering into the glomerular neuropil. Immunoreactive multipolar somata were restricted to the region of the anterior olfactory nucleus and the deep granule cell layer. As in the case of SP, the SS reactive somata appeared to represent only a small subpopulation of cells. Most of the immunoreactive cell bodies extended multiple stained processes. The cross-sectional diameter of the immunoreactive somata ranged from 10 - 17µm (mean = 13.8 + 0.56µm). Thin varicose processes, most likely axons, were followed for short distances from the stained somata or from the anterior olfactory nucleus into the granule cell layer although it was not possible to establish if these were the origin of the immunoreactive processes seen in the more superficial laminae of the olfactory bulb.

Calcium Binding Protein immunoreactivity was limited to a few small neurons that localized predominately to the deep juxtaglomerular zone and less frequently to the external plexiform layer. The somata ranged from 8 to 12µm in diameter (mean = 9.56 + 0.35µm) and typically had a large nucleus surrounded by a relatively thin rim of cytoplasm. Staining of dendritic processes was light but nevertheless suggested that somata in the deep juxtaglomerular zone extended a single tapering dendritic process into glomerular neuropil. Examples in which the immunoreactive processes bifurcated upon entering the glomeruli were not observed. There was no evident staining of axonal processes either extending from the immunoreactive somata or distributed independently within the olfactory bulb.

Parvalbumin immunoreactivity was only rarely encountered in neuronal somata. The few cell bodies that were stained were found at intermediate depths within the glomerular layer and are in accord with the juxtaglomerular designation. Cross sectional diameter of the somata ranged from 11 to 15µm (mean = 12.6 + 0.83µm) with a comparatively large unstained nucleus surrounded by a thin rim of immunoreactive cytoplasm. A complex multiply-branched dendritic arbor, resembling a bush, was often seen extending from the soma into a portion of the neuropil of an adjacent glomerulus. Emerging from the opposite pole of the somas there was frequently a thin process resembling an axon. No stained somata were seen in other layers of the olfactory bulb. However, numerous stained processes resembling axons were apparent in all laminae with the exception of the olfactory nerve layer. These processes were of a uniform non-tapering diameter although regularly spaced and darkly stained varicosities were often present. In the deeper layers of the olfactory bulb, including the presumptive ependymal zone and granule cell layers, the processes appeared to follow a tangent parallel to the surface of the olfactory bulb. In the mitral cell layer and extending into the glomerular layer these processes had an orientation perpendicular to the surface of the olfactory bulb. Individual processes often bifurcated with each branch following a different path. In the glomerular layer the processes distributed within the juxtaglomerular zone extending up to the more superficial aspects of the glomerular layer. No fibers were observed entering into the glomerular neuropil. Similarly, immunoreactive fibers did not extend up into the olfactory nerve layer although they undoubtedly did interdigitate with fascicles of olfactory nerve coursing to more deeply placed glomeruli. It is not yet clear where the majority of these fibers originate; it seems unlikely that the juxtaglomerular neurons immunoreactive for PV are a sufficient source which suggests that centrifugal sources may be involved.

References

Greer, C.A. (1991). Structural organization of the olfactory system. In: T.V. Getchell et al. (Eds.) Smell and Taste in Health and Disease, 65-81. New York, Raven Press.

Smith, R.L., Baker, H., Kolstad, K., Spencer, D. & Greer, C.A. (1991). Localization of tyrosine hydroxylase and olfactory marker protein immunoreactivities in the human and macaque olfactory bulb. Brain Research, 548, 140-148.

Smith, R.L., Baker, H. & Greer, C.A. (1993). Immunohistochemical analyses of the human olfactory bulb. Journal of Comparative Methodology, 333, 519-30.

Smith, R.L. (1992). Immunohistochemical organization of the human olfactory bulb. Unpublished, doctoral dissertation, Yale University School of Medicine, submitted in partial fulfillment of the degree Doctor of Medicine.

The Author
Charles A. Greer

Dr. Charles A. Greer is Associate Professor of Neuroscience and Director of Research in Neurosurgery. His research interests are focused on the olfactory system. The current issues that are being studied in Dr. Greer's laboratory are: 1) the central targets; 2) the synaptic organization of the first central nervous system relay, the olfactory bulb glomerulus; 3) the role of the cytoskeleton in directing mRNAs and cellular organelles to the appropriate subcellular compartments; and 4) continuing studies on the immunocytochemical and synaptic organization of the human olfactory system.

Human Responses to Odour Mixtures: Understanding the Basis for Blending of Odours

David G. Laing
University of Western Sydney

The mechanisms underlying the perception of odorants in mixtures are largely unknown. As part of a larger research programme aimed at understanding the basis of mixture perception, in particular the basis of blending and suppression, the present project investigated the hypothesis that when the olfactory system processes mixtures, different times are required to process different constituents. The hypothesis predicts for example, that odorants which are processed more slowly are likely to be partially or totally suppressed by fast odorants.

To investigate the hypothesis, a unique olfactometer capable of delivering odorants as mixtures or in series separated by time intervals as short as 10 ms was constructed. The results of several experiments with four pairs of odorants showed that:

(i) Mixture constituents are processed at different rates by the olfactory system; time differences of hundreds of milliseconds were recorded.

(ii) Measurement of the time differences between constituents can be achieved using psychophysical techniques with human subjects.

(iii) The type of odorant and its concentration determine which odorant will be perceived first and which is likely to be the one experiencing suppression.

The study clearly demonstrated the existence of temporal coding of odor mixtures and that temporal coding can substantially affect which odors will be discriminated and identified in a mixture and which are likely to be suppressed.

The implications of these findings are far-reaching. Knowledge of temporal differences between odorants opens up a new avenue or tool for the creation of fragrances and flavours. In particular, it should be possible to design and produce fragrances and aromas with specific impact compounds based on their temporal characteristics.

Rationale

The world of the perfumer remains a mystery to the scientist and layperson. Professional perfumers gain their skills and develop their art through years of experience working with odours. They learn how to produce fragrances with specific notes and backgrounds, and in more recent times have benefitted from the additional information provided by sophisticated instruments such as the gas chromatograph and mass spectrometer. Nevertheless, despite the degree of excellence achieved by these artisans, there is no fundamental information and understanding of how they achieve their goals. Clearly, a knowledge of how odorants combine perceptually to produce specific aromas and fragrances would benefit the flavour and fragrance industries and open up new approaches to the development of products.

Over many years, researchers in olfaction have reported examples and conditions where one odour partially or completely suppresses perception of another. In most instances these have been psychophysical studies and have been confined mainly to binary mixtures (Cain & Drexler, 1974; Laing et al., 1984). In this laboratory we have taken these studies one step further and have investigated and observed some of the physiological changes in the nose and olfactory bulb in animals under conditions where suppression of one odour by another occurs in humans (Bell et al., 1987a, 1987b). These studies have been aimed at determining the olfactory centres where perceptual odour interactions occur.

Recently, in our endeavours to understand how and where mixture interactions occur, we attempted to resolve an important fundamental question; how many odours can a human identify in a mixture? Using panels of either untrained or trained laypersons, or professional perfumers and flavourists, we found that the limit was 3-4 odours (Laing & Francis, 1989). In view of the impressive ability of humans to discriminate and identify innumerable odorants when they are presented side by side, the results were dramatic and unexpected. Furthermore, regardless of whether their choice was correct or incorrect, subjects rarely indicated there were more than 3 components present.

Thus, the complexity of mixtures containing 3 odorants was no different to those containing up to 8 odorants. Since most natural and synthetic flavours and fragrances contain dozens even hundreds of components, it can be concluded that very few are perceived and these must be the impact compounds which determine whether a product will be accepted or rejected, liked or disliked.

The loss of information about the constituents of mixtures could be due to one or more mechanisms. First, loss of information could arise if some or all of the cells responsive to one odorant were suppressed by another odorant. This would occur if the suppressor behaved as an antagonist occupying receptor sites normally occupied by the suppressed odorant on the receptor cells in the nose, or by triggering inhibitory circuits in the bulb, or at other olfactory centres. Secondly, loss could occur if the discrete pattern of activity that characterizes stimulation by an odorant in the nose (Astic & Saucier, 1989; Bell et al., 1987b) and bulb (Bell et al., 1987a; Stewart et al., 1979) was combined with those of either odorants in a mixture, into a new pattern at another olfactory centre. Both types of information loss are likely. Evidence for mixture suppression at the olfactory bulb was reported recently by this laboratory (Bell et al., 1987a).

Loss of information through change in activity patterns in the central olfactory system may have its origins in the arrangements of anatomical projections that connect the olfactory bulb to the olfactory cortex. For example, it is well known that there are ordered topographical projections from the nose to the bulb (Astic & Saucier, 1989) and that neural responses to odorants are characterized by discrete patterns of responsive receptor and bulbar cells (Bell et al., 1987a, 1987b; Stewart et al., 1979). However, this order appears to be lost once axons leave the bulb, with projections from the bulb to the cortex exhibiting a low degree of order (Haberly & Bower, 1988). Small areas in the bulbs project to large regions of the cortex and any small region in the cortex samples a broad area in the bulb. This anatomical arrangement suggests that processing at the cortex and beyond would involve a significant degree of combinatorial operation in which axons from scattered sites converge on a target cell. Such convergence would result in loss of any topographical information that characterizes an odour and produce a representation of a complex odour that provides very little information about the constituent odours.

If this is the physiological basis for perceptual blending of odours, how can humans still smell up to 3-4 constituents in a mixture? (Bell et al., 1987a)

One explanation is that odours in a mixture may be discriminated on a temporal basis. Neurophysiological studies have shown that different odours differ greatly in the times they take to stimulate receptor cells, with differences in the order of hundreds of milliseconds being recorded (Getchel et al., 1984). If these temporal differences are maintained at central olfactory structures such as the olfactory cortex, then if only 2-3 odours are present in a mixture, it may be possible for them to be discriminated and identified. However, when more than this number are presented, the time intervals separating the arrival of neural activity from these odorants at the cortex may be too small to allow discrimination, so that only the very fast and slow components, for example, may be discriminated. Identification of the constituents of mixtures, therefore, may be limited by the convergence of neural input at the olfactory cortex but be aided in part by temporal separation of input from each odorant.

Latency differences are also relevant to odour suppression. When one odorant suppresses another, the suppressing odorant may have the shortest stimulating latency, so that not only is it likely to be the first odour perceived, but it may also trigger inhibitory cells and circuits in the bulb and suppress input from the longer latency (slower) odorant.

The proposal that the discrimination of several mixture components is primarily dependent on differences in temporal input is testable and forms the basis of the present study.

Study Design

The overall aim of this research was to determine whether odorants are processed at different rates by the olfactory system and whether this affects their perception in mixtures.

Phase 1: The goal was to construct an olfactometric device for delivering up to 6 odours simultaneously (as mixtures) or in series separated by intervals as short as 10 ms to the nostril of a subject.

Construction of this instrument and subsequent redevelopment has been achieved. The device is shown in Figures 1-3.

Phase 2: The goal was to investigate the effects of delivering pairs of odorants as binary mixtures and in series on the perception of the two constituents. The results indicated that:

(i) The SUPPRESSED odorant in a mixture was always the SLOWER one i.e., the suppressed odorant always required a time advantage when delivered to the nose to overcome the suppression.

(ii) The TYPE of odorant and its CONCENTRATION markedly affected time differences between mixture constituents.

Phase 3: The goal was to determine if it is possible to predict from a knowledge of processing time differences, which odorants will be perceived in mixtures containing up to 6 constituents. Unfortunately, with the time and funds available for this project, it was not possible to pursue this goal. Loss of time occurred primarily as a result of the longer than anticipated time required for construction of the unique olfactometer and for its subsequent re-development to improve its flow and mixing characteristics.

General Methodology

A computer-controlled olfactometer was constructed that was capable of delivering stimuli consisting of between one and six odours as mixtures or in series separated by intervals as short as 10 ms. Specific details of the device are given in Figure 1.

To initiate the delivery of one or more odorants either as a mixture or in series, an instruction from the computer monitor directed a subject to expire into a tube which was sited adjacent to the left nostril, and which contained a pressure-sensitive piezoelectric crystal, then to take a single sniff that ceased at the sounding of a tone 1.5s later. At the onset of the sniff, the pressure change detected by the sensor activated the computer to operate a teflon solenoid valve(s) to allow delivery of an odour(s) to the right nostril of the subject through a parallel tube sited close to the entrance of the latter nostril. All the constituents of the stimulus, therefore, were delivered during the 1.5s trial.

Subjects were administrative and scientific staff at the CSIRO Food Research Laboratory (20 - 55 years of age) who had some experience with sensory testing.

Experiment 1

Aim: To determine if the suppressing odorant in a binary mixture is generally the first odorant to be perceived. Such an observation would support the temporal coding hypothesis.

Methods: The stimuli selected were pairs of odorants whose interactions had been documented previously (Laing, 1988; Laing et al., 1984). For this experiment a single concentration of each odorant

---

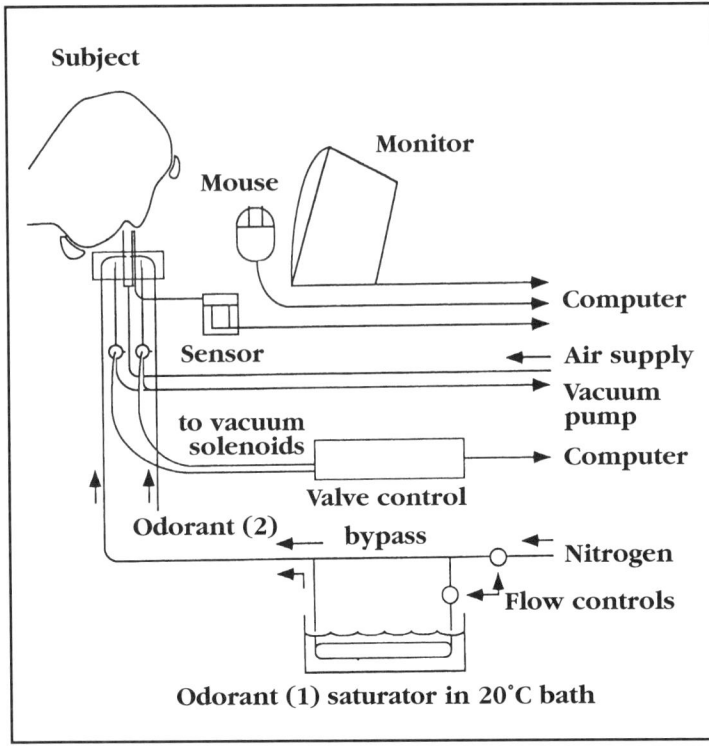

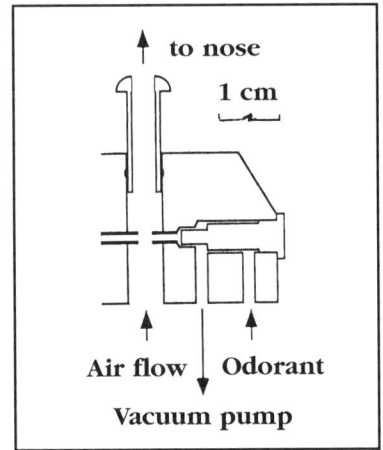

FIGURE 1

Schematic diagram of the olfactometer.

Below is an enlargement of the odour mixing section. Odorant and air flow continuously but the odour stream is exhausted via a vacuum pump between trials.

FIGURE 2

Photograph of a subject sampling the odour stimulus.

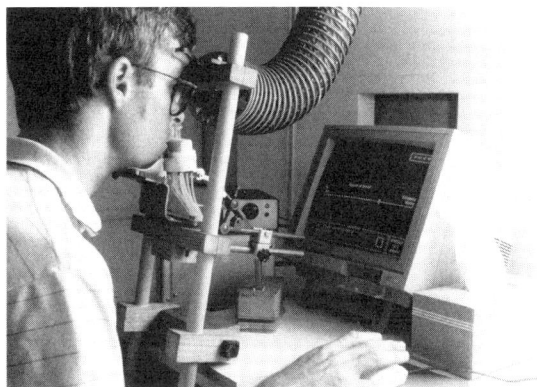

FIGURE 3

Photograph showing the glass saturators containing the odorants in a water bath, flow controllers, and the computer through which the experimenter controls and monitors a subject's performance.

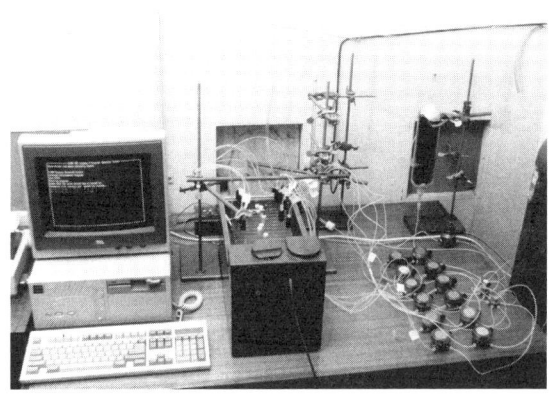

of a pair was selected to represent a condition where one odorant was known to strongly suppress the other.

The odour pairs and respective odour concentrations expressed as the fraction of saturated vapour at 20°C were:

benzaldehyde (.00069SV) - (-)-carvone (.008SV)

(-)-carvone (.0137SV) - (+)-limonene (.045SV)

(-)-carvone (.0125SV) - propionic acid (.00375SV)

(+)-limonene (.009SV) - propionic acid (.00375SV)

All odorants were purchased from Fluka AG, and were of the highest purity available.

With each odour pair the odorants were delivered as a mixture or in series separated by intervals of 13.5, 25, 50, 100, 200 or 400 ms. Each odorant of a pair was given the same time advantage or disadvantage. During each test session each time condition occurred twice so that over the three test sessions 60 responses for each condition were obtained from the panel of 10 subjects. The intertrial interval was 45s. Prior to commencing the experiment subjects participated in several sessions to familiarize themselves with the test procedure and to demonstrate that they could discriminate and

identify each odour with a 90% or better correct response level. After each trial a subject had to indicate which odorant was perceived first. The basic premise being that if an odorant required a time advantage to be perceived first on 50% of trials, this time advantage would correspond to the difference in processing time between two odorants. If the neural input from both odorants is processed simultaneously, no time advantage should be required by either odorant to satisfy the 50% criterion.

Results and Discussion: The results are shown in Figure 4 and Table 1 and are clearcut. With each odour pair the suppressed odorant (indicated by underlining in Table 1) required a substantial time advantage to reach the 50% "perceived first" criterion. The results are significant at the $p < 0.0001$ level, except for the odour pair benzaldehydecarvone where $p = 0.1$.

In brief, the bulk of the results are highly significant but very noisy as indicated by the wide confidence limits. The significance of the "noise" will be discussed later. Importantly, the results support the hypothesis that there are differences between processing times for odorants by the olfactory system and that these affect the perception of the constituents of mixtures.

FIGURE 4

Plots showing the relationship between the percentage of times a stimulus was perceived first and the time separation of the two odorants when they were presented.

The time separation required for an odour to be perceived first on 50% of trials is designated by an arrow (and in Table 1).

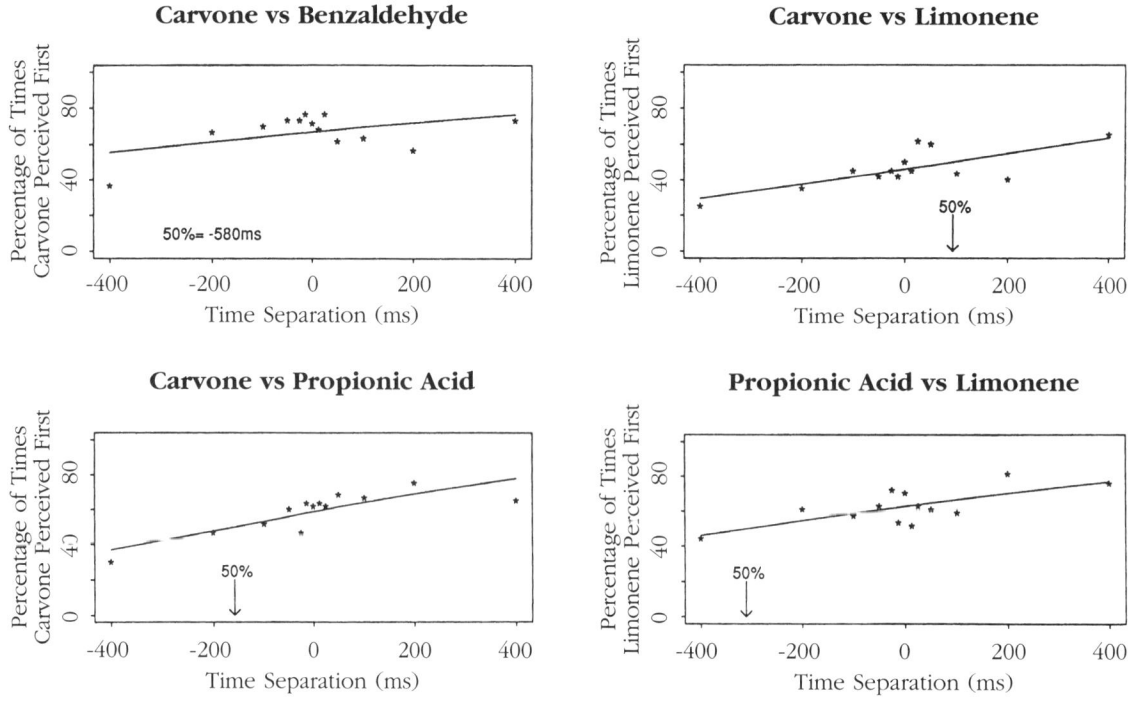

Experiment 2

Aim: To determine if odour concentration affects olfactory processing times.

Methods: Two odour pairs were selected from studies conducted earlier, and there were three concentration levels for each odorant. The concentrations of odorants were adjusted so that for each pair there were three equal intensity levels. The odorants and their concentrations were:

(i) benzaldehyde   (-)-carvone
   (.00056, .00105, .00202SV) (.0038, .0097, .0144SV)

(ii) (+)-limonene   1-butanol
   (.0069, .0110, .0185SV) (.00325, .00613, .01230SV)

All 9 combinations of the odorants of a pair were the stimuli, and each pair was presented as a mixture or physically separated by + 50, 100, 200 or 400 ms. As in the previous experiment subjects were required to identify which odorant was perceived first, and in separate sessions rated the intensity of each odorant when presented alone or in a mixture (zero time advantage).

Results and Discussions: As shown in Tables 2 and 3 a clear and consistent effect of concentration on processing times was recorded. As the intensity of a component in an odour pair was increased there was a corresponding decrease in processing time. For example, with the odour pair benzaldehyde-carvone when both odorants were presented at their lowest concentration/intensity level (level 1),

TABLE 1   **STIMULATION TIME DIFFERENCES**

Odor pair	Perceived first (50%) (ms)	95% Fiducial limits	
		Lower	Upper
Carvone - Limonene	92.0	-8.2	277.2
Carvone - Propionic Acid	159.4	-299.9	-81.8
Limonene - Propionic Acid	311.9	-792.9	-170.1
Carvone - Benzaldehyde	579.6	-	-

Underline signifies time advantage required

TABLE 2   **STIMULATION TIME DIFFERENCES**

Carvone level	Benzaldehyde level	Perceived first (50%) (ms)	95% Fiducial limits	
			Lower	Upper
1	1	+305	51	677
	2	-478	-932	-210
	3	-1230	-2131	-806
2	1	+1305	862	2254
	2	+643	350	1186
	3	-366	-766	-109
3	1	+1350	895	2329
	2	+655	360	1206
	3	-36	-319	232

(+) Signifies time advantage required by benzaldehyde

TABLE 3   **STIMULATION TIME DIFFERENCES**

Limonene level	Butanol level	Perceived first (50%) (ms)	95% Fiducial limits	
			Lower	Upper
1	1,2,3	NO ESTIMATES		
2	1	+590	361	1110
	2	+227	65	510
	3	-597	-1124	-366
3	1	+884	494	2437
	2	+674	359	1873
	3	-133	-522	82

(+) Signifies time advantage required by butanol

benzaldehyde required a time advantage of 305 ms to be perceived first on 50% of trials. As the concentration of benzaldehyde was increased to levels 2 and 3 and that of carvone was maintained at level 1, carvone required time advantages of 478 and 1230 ms to achieve the 50% criterion. Similar changes in processing times were observed for the odour pair limonene-butanol. However, the slope of the regression line for the lowest level of limonene (level 1) with the three concentration levels of butanol was not significantly different from 0. This appears to have occurred because the lowest intensity of limonene was too low for subjects to achieve identification scores greater than 60-70%, unacceptable for a study dependent

on subjects having no difficulty with identification of the target odorants.

Again, as occurred in Experiment 1, with the exception of limonene (level 1) trials, the results were highly significant (p<0.001) but very noisy.

Figure 5 shows the relationship between the perceived intensity of odorants and their concentration when the odorants were presented alone or as mixtures (zero time advantage). This demonstrates that as the perceived intensity of an odorant increases so its suppression of the other member of the pair increases.

For example, (Table 3) as the concentration of carvone increased from level 1 to 3

FIGURE 5

The relationship between the perceived intensity of two odorants in binary mixtures and the level of concentration of each odorant. The three intensity levels in ascending order of limonene are designated (L1, L2, L3); butanol (U1, U2, U3); benzaldehyde (B1, B2, B3); carvone (C1, C2, C3). L0, U0, B0 and C0 indicate conditions where each odorant was at zero concentration. An asterisk indicates that the perceived intensity of an odorant was significantly reduced (p<0.05).

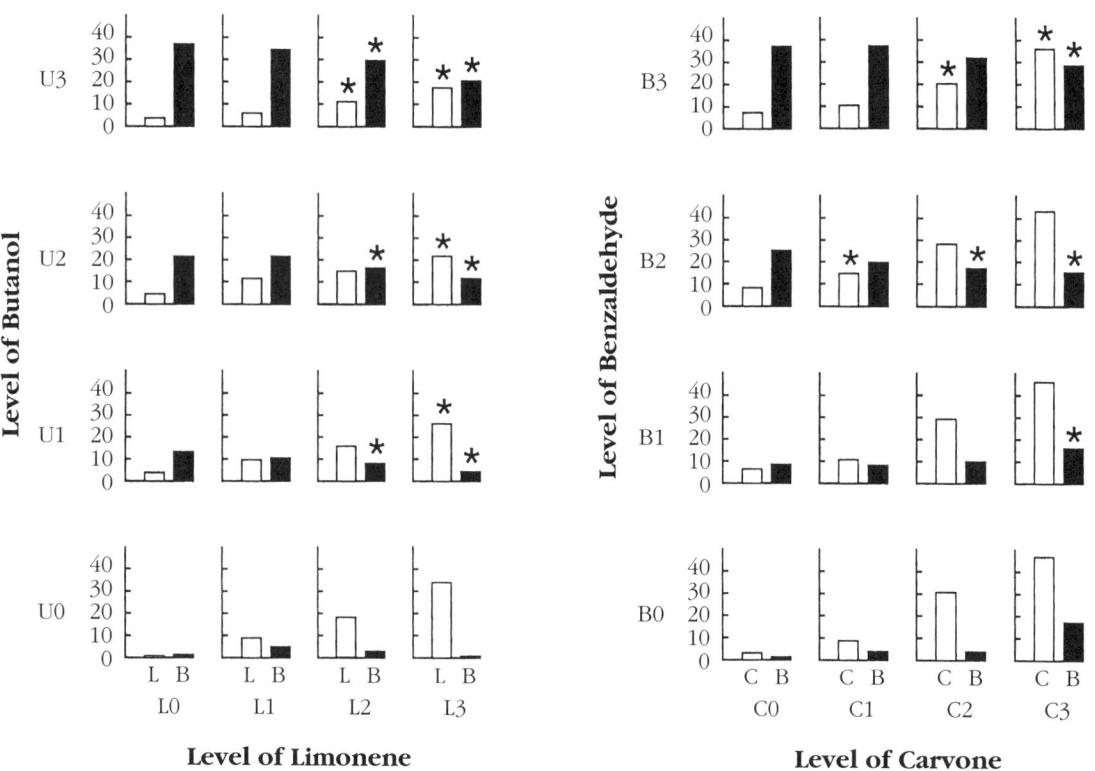

(C1,C2,C3) in mixtrues with the highest concentration of benzaldehyde (B3) the perceived intensity of benzaldehyde decreased from 37.3 to 28.5 (p<0.05). Importantly, changes in the amount and direction of suppression, parallelled changes in processing times. Encouragingly, the direction and degree of suppression exhibited between odorants are similar to those reported earlier (Laing, 1988; Laing et al., 1984).

The results of Experiment 2, therefore, clearly demonstrate that odour concentration markedly affects processing times and influences which odorant of a pair will be perceived first.

The data from Experiment 2 also indicated that processing time differences are also affected by the type of odorant as well as by concentration. For example, with mixtures containing the equal intensity odorants B1C1, B2C2 and L2U2 (Table 4), benzaldehyde and butanol require substantial time advantages (305, 643 and 227 ms respectively) to be perceived first on 50% trials. However, with mixtures of the highest intensity components B3C3 and L3U3 much smaller time differences were recorded, indicating that with higher odour concentrations processing times are much shorter, differences are smaller and at some critical concentration level there may be little difference in processing times regardless of odorant type. This latter condition would apply when the number of odour molecules far

exceeds the number of available receptor sites on receptor cells in the nose.

Finally, consider an explanation for the wide confidence limits or "noise" in the data despite the results in both experiments being highly significant. The present results resemble those reported in studies of vision where subjects were required to indicate the temporal order of perception of 4 single digits presented as a series with intervals of 33 to 100 ms between each digit (Saarinen & Julesz, 1991). In that study, subjects recorded a high level of identification of the individual digits but had great difficulty in indicating the temporal order of perception.

It was proposed that although the digits were processed temporally, the perceptual information was stored in short-term memory whereupon on subsequent retrieval the information on temporal order was largely lost. A similar slow processing mechanism associated with retrieval of olfactory temporal data, therefore, may provide the basis for the unusually high noise in the present data and would be in agreement with the general comment by subjects that the task of describing temporal order was difficult. Thus, it would seem that although odours in mixtures are processed perceptually in a temporal manner, retrieval of the verbal label is slow and interferes with recall of the temporal order of perception.

References

Astic, L. & Saucier, D. (1986). Anatomical mapping of the neuroepithelial projection to the olfactory bulb. Brain Res. Bull., 16, 445-54.

Bell, G.A., Laing, D.G. & Panhuber, H. (1987a). Odour mixture suppression: evidence for a peripheral mechanism in human and rat. Brain Research, 426, 8-18.

Bell, G.A., Laing, D.G., & Panhuber, H. (1987b). Early stage processing of odour mixtures. Ann NY Acad. Sci., 510, 176-7.

Cain, W.S. & Drexler, M. (1974). Scope and evaluation of odour counteraction and masking. Ann NY Acad. Sci., 237, 427-39.

Getchell, T.V., Margolis, F.L. & Getchell, M.L. (1984). Perireceptor and receptor events in vertebrate olfaction. Prog Neurobiol, 23, 317-45.

Haberly, L.B. & Bower, J.M. (1988). Olfactory cortex. Model circuit for study of associative memory. Trends Neurosci, 12, 258-64.

Laing, D.G. (1988). Relationship between the differential adsorption of odorants by the olfactory mucus and their perception in mixtures. Chem Senses, 13, 463-71.

Laing, D.G. & Francis, G.W. (1989). The capacity of humans to identify odours in mixtures. Physiol. Behav., 46, 809-14.

Laing, D.G., Panhuber, H., Willcox, M.E. & Pittman, E.A. (1984). Quality and intensity of binary odour mixtures. Physiol. Behav., 33, 309-19.

Saarinen, J. & Julesz, B. (1991). The speed of attentional shifts in the visual field. Proc. Natl. Sci. USA, 88, 1812-14.

Stewart, W.B., Kauer, J.S. & Shepherd, G.M. (1979). Functional organisation of rat olfactory bulb analysed by the 2-deoxyglucose method. J Comp. Neurol., 185, 715-34.

The Author
David G. Laing

Dr. David G. Laing is the Foundation Professor of Food Technology and Director of the Centre for Advanced Food Research at the University of Western Sydney (Sydney, Australia). He is the author of over 80 journal publications, Senior Editor of two books on the perception of odor and taste mixtures and the human sense of smell and has presented his research work at many international conferences. He leads the Sensory Perception and Development Research Unit at the University and conducts research that is concerned with elucidating the mechanisms of odor mixture perception, the psychophysical and physiological development of the senses of smell and taste in children, and the development of food preferences during early childhood.

Olfactory Perception in Infants: a Retrospective of Research Sponsored by the Olfactory Research Fund

Hilary J. Schmidt
Monell Chemical Senses Center
SUNY - Health Sciences Center at Brooklyn

The research described in this essay was supported by the Olfactory Research Fund and was undertaken to broaden our understanding of the development of olfactory preferences in infants and young children, and to explore the importance of olfaction in early childhood. Initially this research involved developing new methods for assessing reactions to odors in pre-verbal infants, and in children with limited language skills and biases. The development of effective methods has enabled the investigation of two important areas: first, the development of hedonic reactions in infants and young children; and second, the origins of sex differences in responsiveness to fragrances.

Prior research has shown that newborn infants have well developed and functioning olfactory systems that are perhaps more sensitive than those of adults. For example, breast-fed newborns will turn toward the smell of their own mother when it is paired with the smell of another lactating female - an odor discrimination that is particularly difficult for adults (Cernoch & Porter, 1985; Macfarlane, 1975). Although

infants are capable of making fine olfactory discriminations, there is a longstanding controversy concerning whether they can perceive the hedonic quality (pleasantness/ unpleasantness) of odors. Some investigators have argued that adult-like odor preferences and aversions are not apparent until about five years of age. The failure to observe adult-like hedonic reactions prior to the age of five led to the claim that all odor preferences and aversions (including skunk odor) are acquired through experience and associational learning (Engen, 1974; 1982; 1988).

Some of our early work at the Monell Chemical Senses Center challenges the premise that adult-like olfactory preference patterns are absent in pre-school children (Schmidt & Beauchamp, 1988), and raises the possibility that some odor preferences are innate, or at least learned very early on. Indeed the use of a non-verbal, forced-choice task revealed hedonic odor discriminations in 3-year-old children that were remarkably similar to adults' preference patterns. Specifically, when children were asked to indicate liking an odor by pointing to a 'good' puppet (Big Bird) and not liking an odor by pointing to a 'grouchy' puppet (Oscar the Grouch), they categorized a set of 9 different smells (spearmint, floral, wintergreen, strawberry, cloves, banana, vomit, rotten fish, and sweat) in much the same way as adults. We suggested that previous failures to demonstrate hedonic reactions in children less than 4- to 5-years-old may have been due, in part, to methods that were insensitive to the cognitive and behavioral limitations of young children, or to stimulus sets that were too limited in range to capture discriminations (see Schmidt, 1992 for a review).

Our demonstration of hedonic odor discriminations in 3-year-olds challenged the empirical foundations of the radical behaviorist claim that all odor preference patterns are learned. It also led to a search for the origins of olfactory preference patterns in infancy. With the support of the Olfactory Research Fund we went on to develop new methods for assessing olfactory preference patterns in 6- to 9-month-old infants (Schmidt & Beauchamp, 1989). This was an age group previously untested for olfactory perception. Therefore, constructing sensitive and appropriate methods for tapping olfactory perception at this developmental stage presented a unique challenge. Infants 6- to 9-months-old have a wide behavioral repertoire that includes reaching for, exploring, and manipulating objects. We reasoned that if these infants do have hedonic odor reactions, these might be expressed by the ways in which they explored and reacted to odorized objects.

In an initial study (Schmidt, 1990) we video-taped the facial and bodily reactions of 9-month-old babies as they explored in succession (60-90 seconds each) three rattle-like toys which looked alike but had different odors. As judged by adults, one had a pleasant smell (wintergreen), one had an unpleasant smell (spoiled milk), and a third had no smell. Naive adults who viewed the videotaped reactions of the infants were significantly better than chance in judging the hedonic valence of each of the odorized objects, based upon the infant's behaviors as they interacted with each toy. This finding provides direct evidence that infants as

young as 9 months do perceive the hedonic dimension of at least some odors in much the same ways as adults. Careful analysis of a variety of specific reactions including facial expression, however, failed to reveal any single behavior across infants that was associated with a particular hedonic valence. There appear to be large individual differences in the ways that infants expressed their hedonic reactions to these odors. For example, in some infants facial expressions appeared to betray a hedonic reaction, while for others hand gestures, or ways of interacting with the object (e.g., keeping it near or away from the nose) appeared to be a better index of hedonic reaction.

This study provides additional rebuttal to the extreme view that all olfactory preference patterns are learned. An adult-like hedonic dimension of olfactory perception is acquired very early in development, leaving open the possibility that some hedonic odor discriminations are innate. If learning processes do underlie olfactory hedonics, these take place in infancy and do not remain dormant until 3 years of age.

A modified version of this object-odor exploration task was developed to further explore hedonic odor reactions in infancy. In this method we simultaneously (as opposed to sequentially in the previous experiment) offered each infant two objects with different odor properties. In one study we compared how infants responded to an odorized object relative to an non-odorized object; in a second study we compared infants' reactions to two odorized objects - one pleasant and one unpleasant. We reasoned that when offered a simultaneous choice between two toys, infants might spend more time playing with a pleasant smelling toy than one with no smell (experiment 1) or than with a toy with an unpleasant smell (experiment 2). Hence, we were attempting to measure olfactory reactions by the amount of time spent with an odorized object, rather than by evaluating facial and bodily reactions to an object. Since sex differences in responsiveness to odors have been repeatedly demonstrated in adults, we entered sex of the infant into our equation and tested equal numbers of males and females in each of our conditions.

A striking pattern emerged in the results: when offered a choice between an odorized and a non-odorized toy, females spent more time exploring and playing with an odorized toy than a non-odorized toy, even when the odor was unpleasant. In contrast, male babies spent equal amounts of time with the two objects. Since the dependent measure in this study was time spent with an object and not facial expression and bodily movement as in the first study, it remained possible that this method was tapping some aspect of olfactory perception other than hedonic reactions. It was also possible that the contrast between an odorized and a non-odorized object was not appropriate for our purposes: an odorized object may simply be more interesting than an non-odorized object and hence demand more attention. Additionally, the contrast between an odor and no odor may have been an insufficient hedonic contrast to elicit a hedonic choice between the two objects. Finally, the sex difference may have been an artifact of sex differences in general levels of activity and reactivity, and had little if anything to do with olfaction.

With all these considerations in mind, we ran a second experiment in which two odorized objects - one pleasant and one unpleasant - were simultaneously contrasted - a potentially more appropriate test of infants reactions. Again this study yielded a surprising sex difference pattern. This time, female babies explored the two objects equally and displayed a lot of interest in both, while the male babies spent significantly more time with the pleasant smelling (wintergreen) toy. Clearly, these sex effects are mediated by odors, and cannot be explained by a difference in general reactivity. Together, these two studies suggest that sex differences in olfactory processing may have their origins in infancy, and moreover that these sex differences may have a genetic component.

The demonstration of infant sex differences in olfactory processing presents a particular challenge to investigations of the development of olfactory preferences and the mechanisms that may underlie them. The demonstration of hedonic odor reactions in young infants raises questions about the scientific bases of the view that all odor preferences are learned, but it does not tell us how these early olfactory preferences are formed. It remains possible that early experience and learning underlie the development of odor preferences, but that this learning occurs very early in life. Some studies of newborns are, in fact, consistent with this possibility. For example, newborns less than 14-days-old can discriminate the smell of their own mother from the smell of another lactating mother, and they will turn their heads toward an odor to which they have had only a brief exposure: an effect which persists for over two weeks (Davis & Porter, 1991). A recent study suggests that classical learning through association may underlie these effects. Newborn babies who are massaged while an odor is present subsequently prefer that odor (Sullivan et al., 1991).

With these findings in mind, we designed another study to help control for prior olfactory experience, and to explore whether or not exposure to a fragrance in the 9-month-old age range could affect a preference for that fragrance. We reasoned that if olfactory preferences are learned, then an infant should be attracted to a fragrance that has been systematically associated with one of the most positively reinforcing stimuli in the infant's environment - the Mother. Indeed, such an effect has been suggested in newborns (Schleidt & Genzel, 1990). We further reasoned that the sex effects we had observed in our previous studies may have less of an impact on babies' behaviors if we carefully controlled the conditions in which a fragrance was experienced.

In this next study, we had mothers of 9-month-old infants wear one of two perfumes every day for three to five weeks as they went about their daily activities. We then used our object-odor choice task (with the two fragrances embedded in our objects) to assess infants' reactions to both fragrances. If fragance preferences are formed through experience, babies who were exposed to Fragrance 1 on their mothers should prefer it to Fragrance 2, but babies who were exposed to Fragrance 2 should prefer it to Fragrance 1. A great deal of thought, consultation with professional perfumers and extensive pilot testing resulted in the selection of two perfumes which were not likely to be familiar to the infants (none of the mothers had heard of either perfume, and none were in the habit of wearing perfumes regularly), that were clearly discriminable, that were about equally intense when worn, about equally pleasant, and that would last over the course of the day about equally well. One was a woody oriental (FRAGRANCE 1) and another which was predominantly a white flower floriental (FRAGRANCE 2). Half the mothers used FRAGRANCE 1, the others used FRAGRANCE 2. At the end of the 3 to 5-week period, infants were brought into the lab for a testing session. During the testing session infants were familiarized with the two scented rattle-like toys: one scented with FRAGRANCE 1, the other with FRAGRANCE 2. As in the studies described earlier, the infants were offered a simultaneous choice between the two fragranced toys, and their reactions were videotaped.

In contrast to what we would have expected if learning and experience were influencing olfactory preferences, we found that the fragrance exposure had little effect on infants' preferences. Again we observed a sex effect: male infants spent significantly more time with the white floriental, while females spent about the same amount of time with each.

What do these findings suggest? First they suggest that for male infants at least, hedonic preferences amongst fragrances are clear, well formed and not easily modified through classical learning mechanisms by 9 months of age. They can be manifest directly through facial and bodily reactions as in our first study, and indirectly by the choices they make between two odorized objects. Furthermore, the findings suggest that male infants' hedonic discriminations are not limited to extreme odor contrasts between pleasant and unpleasant odors, as we demonstrated in the earlier study, but that they are evident between two fragrances that are adult rated pleasant fragrances. Why do the females fail to respond as males do in this choice task, but evidence clear hedonic reactions when measured directly? These repeated demonstrations of sex differences in responsiveness to odors may reflect a genetic predisposition for females to pay attention to and explore olfactory stimuli to a greater extent than males. Clearly more research is needed to define the nature and conditions under which such olfactory sex differences express themselves.

In sum, the research that has been funded by the Olfactory Research Fund has broadened our understanding of the development of hedonic reactions to odors, and has revealed sex differences in olfactory processing in infancy. The demonstration of hedonic odor reactions in infants demands reconsideration of the view that all olfactory preference patterns are learned relatively late in the pre-school period. The demonstration of sex differences in olfactory processing in infancy must be accounted for in any general explanation of sex differences in adults. Finally, the development of new methods for the study of olfactory perception in infancy serve as tools for the continued study of olfaction in infancy.

Note

Much of this research has been described elsewhere as indicated in the references, and all of it was conducted at the Monell Chemical Senses Center, Philadelphia, in collaboration with Dr. Gary Beauchamp; Ms. Donna Caldwell, Ms. Kathy Chen, Ms. Caryn Cohen all provided invaluable assistance in recruiting mothers and babies, in running the studies and analyzing data.

References

Balogh, R.D., & Porter, R.H. (1986). Olfactory preferences resulting from mere exposure in human neonates. Infant Behavior and Development, 9, 395-401.

Cernoch, J.M., & Porter, R.H. (1985). Recognition of maternal axillary odors by infants. Child Development, 56, 1593-8.

Davis, L.B. & Porter, R.H. (1991). Persistent effects of early odor exposure on human neonates. Chemical Senses, 16, 169-74.

Engen, T. (1974). Method and theory in the study of odor preferences. In Turk, A., Johnson, J., Moulton, D. (Eds.), Human responses to environmental odors, 121-41. New York, Academic Press.

Engen, T. (1982). The perception of odors. New York, Academic Press.

Engen, T. (1988) The acquisition of odour hedonics. In: Van Toller, S. & Dodd, G.H. (Eds.) Perfumery: The psychology and biology of fragrance, 79-90. Chapman & Hill: London, N.Y.

Macfarlane, A.J. (1975). Olfaction in the development of social preferences in the human neonate. Ciba Found. Symp., 33, 103-17.

Makin, J.W. & Porter, R.H. (1989). Attractiveness of lactating females' breast odors to neonates. Child Development, 60, 803-10.

Russell, M.J. (1976). Human olfactory communication. Nature, 260, 520-2.

Schleidt, M. & Genzel, (1990). The significance of mother's perfume for infants in the first weeks of life. Ethology and Sociobiology, 11, 145-50.

Schmidt, H.J. & Beauchamp, G.K. (1988). Adult-like odor preferences and aversions in three-year-old children. Child Development, 59, 1136-43.

Schmidt, H.J. & Beauchamp, G.K. (1989). Sex differences in responsiveness to odors in 9-month-old infants. 11th Annual meeting of Achems XI, Sarasota, FL.

Schmidt, H.J. (1990). Adult-like hedonic responses to odors in 9-month-old infants. Achems XII, Sarasota, FL.

Schmidt, H.J. (1992). Olfactory Hedonics in Infants and Young Children. In Van Toller, S. & Dodd, G.H. (Eds.) Fragrance: The psychology and biology of perfume, 27-35. Elsevier, Cambridge.

Sullivan, R.M., Taborsky-Barba, S., Mendoza, R., Itano, A., Leon, M., Cottman, C.W., Payne, T.F. & Lott, I. (1991). Olfactory classical conditioning in neonates. Pediatrics, 87, 511-18.

The Author
Hilary J. Schmidt

Dr. Hilary Schmidt received a B.A. from the University of British Columbia, and a Ph.D. in psychology in 1985 from the University of Pennsylvania. Her dissertation research focused on visual perception in infants and young children. She then proceded to the Monell Chemical Senses Center for a post-doctoral fellowship, where she initiated research on olfactory perception in infants and pre-schoolers. There she developed several new and innovative methods for studying hedonic reactions to odorants suitable for children and babies with limited or no verbal skills. Her research at Monell culminated in several journal articles and a number of book chapters. She has subsequently moved into the field of medical education, and is currently the Assistant Dean for Curriculum Evaluation and Faculty Support and Columbia University Health Sciences.

Twin Research on Olfactory Characteristics

Nancy L. Segal

Department of Psychology and Twin Studies Center
California State University, Fullerton, CA

Studies of monozygotic (MZ or identical) and dizygotic (DZ or fraternal) twins provide a very informative approach to understanding genetic and environmental influences on human physical and behavioral variation. The logic underlying twin research methodology is simple yet elegant. Increased resemblance between MZ co-twins relative to DZ co-twins is consistent with a genetic component underlying the trait of interest. This is because MZ twins share all their genetic inheritance, while DZ twins share half their genetic inheritance, on average, by descent. Thus, while MZ twins may differ only as a function of environmental events, DZ twins may differ due to genetic or environmental causes.

Individual differences in odor identification and sensitivity have been a source of fascination for many years. Age, gender and familiarity are often used to explain observed olfactory variations among people. In contrast, a twin research perspective on individual differences in olfaction, with a view toward identifying a genetic component, has been applied less often. Twin studies in the United States and abroad have demonstrated genetic influence on sensitivity to androstenone and to isoamyl acetate (Grosse-Isseroff etal, 1992; Wysocki & Beauchamp, 1984). A twin study of phenyl ethyl alcohol (PEA) sensitivity, odor identification and odor characterization has recently been completed by investigations at the Twin Studies Center at California State University, Fullerton (CSUF). A discussion of research methods and findings of that study are presented here.

Odor Sensitivity, Identification, and Characterization

The sample was composed of 83 adolescent and adult twin pairs, recruited through notices in local newspapers, Mothers of Twins Clubs and personal referrals. Twins ranged in age between 10.9 and 82.7 years. Zygosity diagnosis (classification of twins as MZ or DZ) was accomplished

FIGURE 1

Identical female twins, Pair 1.

FIGURE 3

Identical female twins, Pair 2.

FIGURE 2

Dr. Nancy L. Segal, Director of the Twin Studies Center, administering the PEA detection test to one pair member

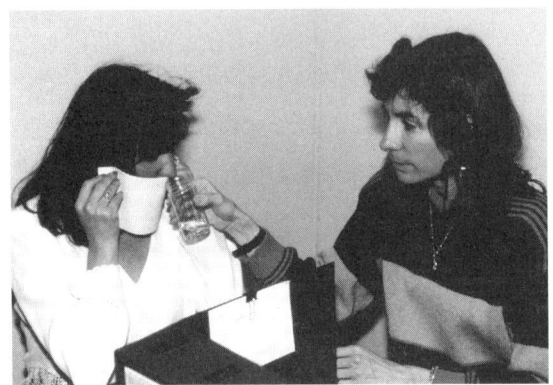

FIGURE 4

Dr. Kathleen W. Brown administering the PEA detection test to one pair member

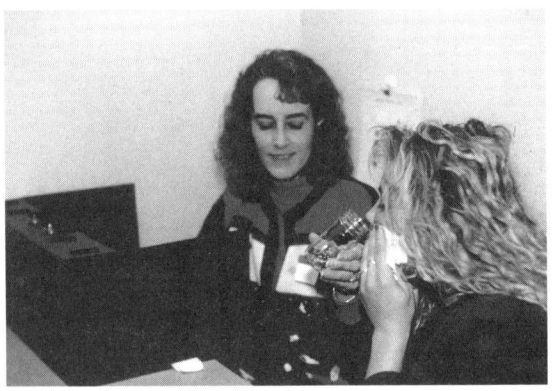

by extensive blood-typing analysis or physical resemblance questionnaires. The final sample included 46 MZ twin pairs and 37 DZ twin pairs; 11 DZ twin pairs were opposite-sex. All olfactory tests were administered to pair members by different examiners to eliminate bias associated with knowledge of twin type. Figures 1 through 4 illustrate this procedure with two pairs of identical female twins.

Sensitivity to PEA was assessed by a sensory detection threshold test. Twins were presented with pairs of bottles containing differing concentrations of PEA. A forced choice, single staircase procedure, as described by Doty,

Shaman & Dann (1984), was used. Twins were instructed to sniff the bottles and to identify the bottle which contained the "stronger" odor. By varying the concentrations over a series of tests, the level at which each individual detected the scent was calculated. Genetic influence on PEA sensitivity was not suggested, given the similar correlations found for MZ and DZ twin pairs. MZ male twins were somewhat more similar than MZ female twins, but the difference was modest. The small number of available DZ male twin pairs did not allow a similar comparison for DZ twins.

The University of Pennsylvania Smell Identification Test (UPSIT) is a scratch and sniff test which requires approximately 15 minutes for completion (Doty, et al., 1984). A pencil is used to scratch a label which releases an odor. Twins were instructed to choose the item (from a list of four possible items) that most closely matched their smell experience. A score on the UPSIT is equal to the number of odorants correctly identified out of 40. Following odor identification, twins rated each odor with respect to intensity, pleasantness, coolness, irritation and familiarity.

Females scored significantly higher than males on the UPSIT, a finding consistent with many previous analyses of gender differences in olfaction. Greater resemblance between MZ male twins than between MZ female twins suggested that genetic influences may differentially affect odor identification in males and females. DZ female twins were less alike than MZ female twins but, unfortunately, there were too few DZ male twin participants to enable a meaningful comparison. Opposite-sex twins showed an absence of resemblance on this task. Collectively, these findings highlight the importance of examining olfactory data separately for males and females.

Henning's (1915) classification system was used to organize the 40 UPSIT odors into six categories (fruity, spicy, flowery, resinous, foul and burnt). Genetic effects were suggested for rating of the Weak-Strong dimension for three of the odor classifications (spicy, flowery and burnt). Ratings on the dimensions of Unpleasant-Pleasant, Nonirritating-Irritating and Unfamiliar-Familiar for the Flowery category also suggested genetic influence. Mechanisms underlying these results warrant further study.

Research in Progress: Twin Studies of Kin Recognition

A number of investigators have addressed the possibility that attraction between individuals and recognition among relatives may be partly based on olfactory cues. Richard H. Porter at the George Peabody College at Vanderbilt University, has suggested (1991) that humans possess "olfactory signatures" (i.e. unique odors produced by individuals) which may be involved in recognition and communication. Therefore, a relationship between genetic similarity and odor similarity would be anticipated. This reasoning suggests that MZ twins should show greater similarity in body odor than DZ twins. Most studies in this area have included mothers and infants, but several twin studies are available. Data collection for a twin study of this type has been completed in the laboratory at CSUF.

MZ and DZ twin adolescents and young adults were asked to wear t-shirts for three consecutive nights. During this time they were instructed to refrain from wearing perfume or other cosmetics, and were asked to wash only with Ivory soap. T-shirts were stored in sterile plastic bags during the day to prevent contamination by odors in the environment. Participants were also asked to complete a diet diary while participating in this phase of the study. A panel of judges then sniffed a particular t-shirt and selected the "relative" from an array of three t-shirts. (These three shirts belonged to the co-twin and to two unrelated twins of the same age and sex.) The level of confidence that judges associated with each of their ratings were also requested. It is anticipated that greater accuracy in matching the t-shirts of MZ twins than DZ twins will be observed. This finding would support a genetic influence on body odor. Contrary to expectation, the proportion of correct identifications for MZ twins did not significantly exceed that of DZ twins. A lack of relatedness and/or familiarity of the judges with the twins may be associated with the non-significant findings. Most previous studies which have demonstrated accuracy in identification of individuals by olfactory cues from garments have used genetically related or familiar individuals as judges. A significant negative relationship between twins' age and proportion of accuracy perpair was observed ($r=-.42$, $p=.01$, $n=37$).

Summary

A genetic perspective has become increasingly apparent in many domains of human behavioral and physical functioning. It appears likely that twin studies will continue to play a significant role in clarifying the genetic and environmental influences underlying individual differences in olfactory characteristics. Studies of adoptive parent-child pairs and adoptive sibling pairs (who share environments, but not genes) would further enhance understanding of normal and abnormal olfactory functions.

A more comprehensive report of the present research is available in Physiology and Behavior, 57, 605-9.

References & Additional Readings

Doty, R.L., Shaman, P. & Dann, M.S. (1984). Development of the University of Pennsylvania Smell Identification Test, a standardized microencapsulated test of olfactory function. Physiology and Behavior, 32, 489-502.

Gross-Isseroff, R., Ophir, D., Bartana, A., Voet, H., & Lancet, D. (1992). Evidence for genetic determination in human twins of olfactory thresholds for a standard odorant. Neuroscience Letters, 141, 115-8.

Porter, R.H. (1991). Human reproduction and the mother-infant relationship: The role of odors. Smell and Taste in Health and Disease. In T.V. Getchell, R.L., Doty, L.M. Bartoshuk, and J.B. Snow, Jr. (Eds.). Raven Press, New York, 429-42.

Segal, N.L., Brown, K.W. & Topolski, T.D. (1992). A twin study of odor identification and olfactory sensitivity. Acta Genetica Medicae et Gemellologiae, 41, 113-21.

Segal, N.L. & Topolski, T.D. (1995) The genetics of olfactory perception. In R.L. Doty (Ed.) Handbook of Clinical Olfaction and Gustation, 323-43. New York: Marcel Dekker, Inc.

Segal, N.L., Topolski, T.D., Wilson, S.M., Brown, K.W. & Araki, L. Twin analysis of odor identification and perception. Physiology and Behavior, 57, 605-9.

Wysocki, C.J., & Beauchamp, G.K. (1984). Ability to smell androstenone is genetically determined. Proceedings of the National Academy of Sciences, 81, 4899-902.

The Author
Nancy L. Segal

Dr. Nancy L. Segal is a Professor of Developmental Psychology and Director of the Twin Studies Center at California State University, Fullerton. She is a Fellow in both the American Psychological Association and American Psychological Society. Dr. Segal is also Contributing Research Editor for Twins Magazine and a member of the Advisory Board for the Center for Loss in Multiple Birth.

Fragrance and Psychophysiology

II

People enjoy fragrance for its purely aesthetic value, but like other forms of sensory stimulation, fragrance seems to provide something beyond beauty. To encourage exploration of these additional effects, the Olfactory Research Fund has sponsored projects that examine how fragrance affects such basic psychophysiological processes as sleep, attention, and anxiety.

Are we responsive to odors while we sleep? <u>Pietro Badia, Michelle Boecker and Kenneth Wright, Jr.</u> recorded brain wave activity and other measures to determine how fragrance delivered during sleep alters the well-known physiological patterns of sleep.

People describe some scents as "refreshing," and this may be more than a metaphor. Attention level can be measured in the laboratory by means of a vigilance test. <u>William Dember, Joel Warm and Raja Parasuraman</u> used this technique along with fragrance stimulation and found that certain scents are effective in maintaining sustained attention.

We often select scents and other forms of sensory stimulation to set a mood and help us relax. These everyday practices suggest a potential role for fragrance in reducing the anxiety caused by unusually stressful conditions. <u>William Redd and Sharon Manne</u> examined patients undergoing magnetic resonance imaging (MRI), and found that providing fragrance during the procedure significantly reduced patient anxiety levels.

Some Effects of Fragrances on Sleep

Peter Badia
Michelle R. Boecker
Kenneth P. Wright, Jr.
Bowling Green State University

Over the past several years we have studied the effects of fragrances on initiating and maintaining sleep. Our early research documented changes in brain waves, heart rate and behavior that occurred when odors were presented during sleep. Subsequent studies suggested that odors judged relaxing or alerting during waking might have differential effects on sleep quality. Most of these studies indicated that odors presented prior to or during sleep were alerting. Recently we focused on the sleep effects of heliotropin (an extract of the valerian herb), androstenone (a reputed sex pheromone with a musky odor) and Galaxolide (a synthetic musk). Additionally, we were interested in whether odors presented during rapid eye movement sleep (dreaming, sleep) would affect the emotional character of the dream, and whether they would be incorporated into the dream. Some of this research is described below.

Experiment 1 – The Effects of Heliotropin on Sleep.

We hypothesized that odors judged relaxing during waking might enhance sleep and those judged alerting during waking might degrade sleep. Earlier research reported findings consistent with this hypothesis (Badia et al., 1990). The odors of jasmine or peppermint were judged alerting, and when presented prior to or during sleep tended to disrupt sleep architecture and sleep continuity. On the other hand, the odor of heliotropin (judged relaxing) tended to enhance sleep in one study, and to disrupt sleep in another. The latter studies were conducted using relatively short daytime naps.

We then decided to address the contradictory findings with heliotropin by using a full night of polygraphically recorded sleep.

Subjects and Apparatus

Subjects were 18 female undergraduate students at Bowling Green State University. They were screened prior to participation for medication use and for sleep, health and allergy problems.

Grass 7P511 AC amplifiers (Model 78D polygraph) were used to record brain wave activity (EEG), eye movement activity (EOG) and chin muscle activity (EMG). White noise (65 dB SPL) was generated by a Grason Stadler 455B Noise Generator.

The odor delivery system consisted of a Whisper 1000 aquarium pump, charcoal air filter, glass bottles (35 ml), air flow control valves, TYGON (7 mm) and Teflon (2 mm) tubing, and an oxygen mask (B & F Medical-Model 64009 Pediatric Oxygen Mask). Both the fragrance and unscented room air were delivered via the mask. The heliotropin odor was embedded in 2mm pellets of low-density polyethylene plastic. Odor concentration at the mask was controlled by the air flow and the number of pellets used. One set of valves allowed the presentation of unscented, filtered, room air, and another set allowed the presentation of heliotropin. To prevent odor contamination, heliotropin and filtered room air were presented through separate Teflon tubes. A constant flow of 0.2 liter/min (odor or room air) was presented through the mask from lights out until the subject awakened.

Procedure

Following their arrival to the Sleep and Psychophysiology Laboratory, subjects were given a description of the procedure and informed consent was obtained. They were randomly assigned to receive either the heliotropin odor or unscented filtered room air. Electrodes were secured for recording EOG, EMG, and EEG (sites C3, O1). Subjects were escorted into an electrically shielded, sound-attenuated bedroom where they slept for approximately 8 hours. Sleep records were scored for sleep efficiency, sleep staging, latency to sleep stages and persistent sleep (10 minutes of uninterrupted sleep), and number of spontaneous awakenings and stage shifts.

Results

Data were analyzed using Analysis of Variance (ANOVA) techniques. Most sleep measures showed little difference between the heliotropin and filtered room air conditions. However, a significant difference in the number of stage shifts was observed. More stage shifts occurred under the heliotropin condition compared to the filtered room air, suggesting that the odor may fragment sleep. There was no evidence that heliotropin enhanced sleep.

Experiment 2 – The Effects of Androstenone on Sleep.

This study assessed the effects of the odor of androstenone on sleep. Perception of androstenone is highly variable among individuals. Some individuals describe it as having a pleasant musky smell, while others describe it as unpleasant and resembling stale urine or strong sweat (Wysocki & Beauchamp, 1984). Androstenone has a pheromonal function in pigs, and it has been suggested that androstenone may play a similar role in human sexual behavior (Filsinger, et al., 1984; Gustavson, et al., 1987). The odor has been shown to influence choice behavior and impressions of self and others (Filsinger, et al., 1984; Gustavson, et al., 1987). An anosmia specific for androstenone has been reported: approximately 47% of people are unable to detect it (Wysocki & Beauchamp, 1984). The present experiment examined the effects of androstenone on sleep using one-hour naps. Subjects were 65 undergraduate students at Bowling Green State University. They were screened prior to participation for medication use, sleep habits, health and allergy problems, and anosmia for specific odors. Androstenone (15 ml in crystalline form) was dissolved into 25 ml of white, odorless mineral oil, as described in the previous experiment. All subjects were polygraphically recorded. Sleep records were scored for sleep efficiency, sleep staging, latency to sleep stages, and persistent sleep and number of spontaneous awakenings and stage shifts.

Results

The presentation of androstenone tended to disrupt sleep on most recorded sleep measures. Total sleep time and sleep efficiency decreased and awakenings increased for the androstenone condition compared to room air alone. Latency to persistent sleep, latency to stages 1, 2, 3/4 and to REM were all longer for the androstenone condition. Additionally, the androstenone condition showed a higher percentage of wake time, and a lower percentage of stage 3/4 and REM sleep compared to the room air condition. These results are consistent with our previous findings that androstenone tends to be disruptive when presented during sleep.

We then assessed the effects of androstenone on the sleep of anosmics (subjects unable to smell the odor) and osmics (subjects able to smell the odor). We were also interested in whether the hedonic rating of androstenone (as pleasant or unpleasant) had an effect on sleep.

Subjects in the androstenone condition were divided into 4 groups, osmic subjects rating the odor as pleasant, unpleasant, or neutral, and those anosmic to the odor. Several analyses were performed.

The first analysis assessed whether subjects (osmic and anosmic) receiving the odor differed from those receiving unscented filtered, room air. The data showed that anosmics and osmics were similar to each other but tended to differ from control subjects (those receiving only filtered room air) on most recorded sleep measures. Total sleep time for anosmic (26.89 min) and osmic (28.38 min) subjects receiving androstenone was reduced relative to subjects receiving filtered room air (34.55 min). Similarly,

sleep efficiency was reduced for anosmic (44.77%) and osmic subjects (47.31%) relative to control subjects (57.57%). Additionally, latency to persistent sleep, stage 2 and stage 3/4 sleep tended to be longer for the anosmic and osmic groups relative to filtered air controls. Finally, anosmic and osmic subjects tended to spend more time in wake and less time in stage 3/4 sleep relative to controls. These findings suggest that androstenone is slightly disruptive to sleep, regardless of whether an individual can smell the odor or not.

To assess whether the effects of androstenone on sleep varied depending on the subject's hedonic perception of the odor, a second analysis compared subjects rating the odor as pleasant, unpleasant or neutral with those subjects receiving unscented, filtered room air. This analysis showed that the odor tended to disrupt sleep of subjects rating the odor as either pleasant or unpleasant relative to control subjects receiving filtered room air. Those rating androstenone as neutral did not differ from those receiving filtered room air. Given the latter findings, subjects rating the odor as either pleasant or unpleasant were combined and compared to those receiving filtered, room air; those rating the odor as neutral were excluded. These analyses revealed that androstenone was disruptive on all 12 recorded sleep measures when compared to filtered, room air. A significantly greater number of stage shifts, awakenings, longer latency to stage 3/4 sleep, and reduced stage 3/4 sleep were observed for the androstenone condition. Thus androstenone appears to disrupt sleep when it is perceived as pleasant or unpleasant, but not when it is rated as neutral.

Experiment 3 – The Effects of Galaxolide on Sleep.

We were interested in whether musky odors in general were disruptive to sleep, and whether a synthetic musk such as Galaxolide would have effects similar to androstenone. Research with Galaxolide during waking has shown that this musky odor impairs performance (Lorig et al., 1989).

Subjects were 20 female undergraduate students at Bowling Green State University. They were screened prior to participation for medication use, and for sleep, health and allergy problems. All subjects were polygraphically recorded, as described in the previous experiments.

Results

The differences between Galaxolide and room air were small and inconsistent compared to the previous findings with androstenone. This was the case for all important measures including total sleep time, sleep efficiency, latency to stage 1, 2 and 3/4 sleep, and time spent awake. Thus, in contrast to the androstenone findings, Galaxolide did not disrupt sleep.

Experiment 4 – Odors Presented during Rapid Eye Movement Sleep (REM).

Research in visual, auditory and somatosensory psychophysiology has shown that stimuli presented during sleep can be detected and registered by EEG as well as other measures. The purpose of this experiment was to determine whether odors (androstenone and peppermint) presented during REM sleep 1) altered the EEG waves, 2) were incorporated into the dreams, and 3) affected the emotional content of dreams. Subjects were undergraduate students (6 male and 14 female) at Bowling Green State University. They were screened prior to participation for medication use, and for sleep, health and allergy problems. All subjects were polygraphically recorded as described in the previous experiments.

Procedure

Subjects received either an odor or filtered room air whenever the EEG and EMG records indicated the occurrence of REM sleep. Ten subjects received alternating presentations of peppermint and room air, and 10 others received alternating presentations of androstenone and room air. Odor or room air was presented at the start of each REM period and continued for 15 minutes. At the end of odor presentation, the subject was awakened and asked to give a dream content report, including emotional content of the dream. The different sense modalities (olfaction, audition, tactile, etc.) experienced in the dream were recorded.

EEG was sampled at 128 Hz. Spectral power analysis in the delta, theta, alpha, beta1 and beta2 bands was performed for each recording site. In addition, facial electromyography was used to examine the effects of odors on dream affect. In waking, facial muscle activity has been used as an indication of emotional experience (6,7). A recent study showed that the facial musculature is active during REM sleep in some vivid dreamers (5). Therefore, facial electromygraphic activity was recorded from corrugator and zygomatic muscle sites during odor and room air control trials.

Results

Odor vs. Air. No significant EEG differences in spectral power were observed between odor and room air presentations for the Cz and Pz scalp sites.

Differences between fragrances. EEG spectral power differences between androstenone and peppermint were analyzed for both brain sites and across all five EEG bands. There was significantly more EEG power in the theta band during androstenone trials compared to peppermint trials at the Cz and Pz brain sites.

Emotionality. An analysis determined that the emotional content of the dreams was different for odor vs. room air presentations. "Happiness" tended to be higher and "disgust" tended to be lower for both odors relative to room air. There were no differences in self reported "happiness" and "disgust" between androstenone and peppermint. None of the other rated emotions were significantly different between the groups.

Facial EMG. We expected greater zygomatic activity and less corrugator activity on odor versus air trials given that there was differential emotionality. Contrary to expectations, no significant differences in muscle activity were found between odor and air trials. Nor were there differences between odors.

<u>Incorporation into Dreams</u>. Incorporation of some aspect of the experimental situation (technician, equipment, testing, electrodes, etc.) was found on 19% of the reported dreams. This suggests that stimuli derived from the laboratory environment were far more salient than the odors presented during REM sleep, where the incorporation rate for the odors was low, approximately 4%.

Experiment 5 – Odors presented during REM.

Study 5 is an extension and partial replication of the previous study. The scalp sites selected for our first study (Cz and Pz) were based upon awake EEG findings. However, recent research using PET scans suggests that the right side of the brain, specifically the frontal region, may be more active during REM sleep. Therefore, the next study focused on right hemisphere EEG (F4, C4, T4, O2). The right hemisphere has also been purported to be involved in emotion and may thus reveal information about emotional responses to the odors during dreaming. An additional aim of the present study was to determine whether the previous findings could be replicated. We were also interested in determining whether an odor rated more negatively would produce EEG and mood effects different from the more positively rated odors. In this regard, we tested the odor lavandin grosso which is rated more negatively than the odors used previously. Subjects were 7 male and 17 female undergraduate students at Bowling Green State University. They were screened prior to participation for medication use, and for sleep, health and allergy problems.

Procedure

See procedure section from Experiment #4 for details. Eight subjects received peppermint and room air, 8 subjects received androstenone and room air, and 8 subjects received lavandin grosso and room air. Odor or room air was presented at the start of each detectable REM period for 5 minutes. After 5 minutes of odor presentation subjects were awakened and asked to give a dream report, including the emotional content of the dream. If a stage change or awakening occurred before the end of the 5 min period the odor was turned off and subjects were asked to give a dream report.

Results

<u>Spectral EEG Data</u>. Differences between odor and room air presentations were observed for lavandin grosso. There was significantly more power in the alpha band for brain sites C4 and O2 (and a trend at brain site F4) for the fragrance trials compared to room air control trials. In addition, there were trends for lavandin grosso to have more spectral EEG power in delta and beta1 at brain site O2 during fragrance trials when compared to room air trials. It was also apparent that lavandin grosso affected the brain wave patterns when presented during REM sleep since EEG activity was much more variable. In contrast, the differences for androstenone and peppermint compared to filtered room air were small.

<u>Differences between fragrances</u>. In general, significant differences in EEG power were seen when comparing lavandin grosso to either androstenone or peppermint fragrance. There was always more power in the EEG during lavandin grosso trials compared to androstenone or peppermint trials. This suggests that the brain is being affected by this odor. There were very few differences between androstenone and peppermint, thus replicating the findings of our last study.

<u>Emotionality</u>. Contrary to expectation, no significant differences were found for emotionality of the dream when an odor was presented compared to room air control trials, thus failing to replicate findings of the previous study. The differences were small for the self-reported emotions between androstenone, peppermint or lavandin grosso and room air only. There were no significant differences between the three odors for any of the self-reported emotions during stage REM.

<u>Odor Incorporation.</u> As in the preceding study, 15% of the dreams reported in REM sleep showed incorporation of the experimental situation. This replicates our earlier finding, and again suggests that the laboratory and experimental situation were more salient to the subject than the odor presented during REM sleep.

General Conclusion

Our research shows that most odors presented prior to or during sleep tend to disrupt normal sleep architecture, staging and continuity. No odor tested had an enhancing effect on sleep architecture or sleep continuity. Even odors judged as relaxing during wakefulness were disruptive to sleep. The results of the androstenone study indicate that androstenone is disruptive to sleep regardless of an individual's ability to smell it. The fact that subjects anosmic to the odor tended to be affected by it suggests that stimulation of the olfactory receptors occurred and the brain was receiving stimulation; and/or that other aspects of the olfactory system were affected by the odor. Regarding the latter, a recent study suggests that the vomeronasal organ is functional in humans and may provide the basis for stimulating the sleeping brain since it acts as a pheromone detection system in animals (8). The finding that the sleep of subjects anosmic to androstenone was disrupted supports the finding that androstenone can have behavioral effects even though the odor cannot be detected (anosmic subjects). Whether this finding is related to androstenones' purported role as a pheromone is unknown. Other odors with specific anosmias should be tested in this regard. It was also interesting that androstenone was more disruptive to sleep for subjects rating the odor pleasant or unpleasant compared to neutral. The latter suggests that an odor with a strong hedonic rating, regardless of whether it is pleasant or unpleasant, may disrupt sleep.

References

Badia, P., Boecker, M., & Lammers, W. (1990). Some effects of different olfactory stimuli on sleep. <u>Sleep Research</u>, <u>19</u>, 145.

Filsinger, E., Braun, J., Monte, W., & Linder, D. (1984). Brief communication. <u>Journal of Comparative Psychology</u>, <u>98</u>, 219-22.

Gustavson, A., Dawson, M., & Bonett, D. (1987). Androstenol, a putative human pheromone, affects human (homo sapiens) male choice performance. <u>Journal of Comparative Psychology</u>, <u>101</u>, 210-12.

Lorig, T., Huffman, E., DeMartino, A., & DeMaro, J. (1989). EEG effects of low level galaxolide administration. <u>Chemical Senses</u>, <u>14</u>, 281-84.

Perlis, M.P., Wright, K.P., & Bootzin, R.R. (1990). Sustained facial muscle activity during REM sleep. <u>Sleep Research</u>, <u>19</u>, 141.

Schwartz, G.E., Fair, P.L., Salt, P., Mandel, M.R., & Klerman, G.L. (1976). Facial muscle patterning to affective imagery in depressed and nondepressed subjects. <u>Science</u>, <u>192</u>, 489-91.

Schwartz, G.E., Fair, P.L., Salt, P., Mandel, M.R., & Klerman, G.L. (1976). Facial expression and imagery in depression: An electromyographic study. <u>Psychosomatic Medicine</u>, <u>38</u>, 337-47.

Stensaas, L., Lavker., R., Monti-Bloch, L., Gosser, B., & Berliner, D. (1991). Ultrastructure of the human vomeronasal organ. <u>J. Steroid Biochem. Molec. Biol.</u>, <u>39</u>, 553-60.

Wysocki, C., & Beauchamp, G. (1984). Ability to smell androstenone is genetically determined. <u>Proc. Natl. Acad. Sci. USA</u>, <u>81</u>, 4899-902.

The Author
Peter Badia

Dr. Peter Badia holds the title of Distinguished University Professor at Bowling Green University. The focus of his research includes psychophysiology, psychophysiology of sleep and circadian rhythms. Dr. Badia has published over 100 articles in scientific journals and has participated in numerous international conferences.

Olfactory Stimulation and Sustained Attention

Joel S. Warm and William N. Dember

Raja Parasuraman

William N. Dember
Joel S. Warm
University of Cincinnati

Raja Parasuraman
The Catholic University of America

Many claims have been made about the ability of certain fragrances to affect mood states, memory, and performance, but there is still little hard evidence from laboratory research to support such claims. The purpose of the experiments reported here was to investigate the ability of fragrances to enhance performance accuracy on a tedious but demanding vigilance task. Vigilance, or sustained attention tasks, as performed in the laboratory, are meant to simulate the core features of "real-world" tasks engaged in by quality control inspectors, radar operators, and other personnel who must monitor displays for the occasional, unpredictable occurrence of critical events or signals (Warm, 1993). We were motivated in our efforts by the results of self-report studies that certain fragrances, peppermint among them, are "arousing" or "stimulating" as well as by evidence to that effect from electroencephalography (Harver, et al., 1989; Lorig & Schwartz, 1988).

The results of four experiments supported by the Olfactory Research Fund are summarized in this paper. The first of them has been published (Warm, et al., 1991); the other three are being prepared for publication.

1. Effects of olfactory stimulation on performance and stress in a visual sustained attention task.

On the basis of an extensive pilot study of twelve candidate fragrances, two were selected for use in this experiment: peppermint and muguet (lily of the valley). Both fragrances had been found to be very pleasant; in addition, subjects rated the peppermint fragrance as stimulating or arousing, and they rated muguet as highly relaxing. Given those self-report data, which closely matched results obtained by the supplier, International Flavors and Fragrances, we expected that peppermint would directly facilitate performance accuracy by helping subjects maintain their level of arousal. By virtue of its relaxing ratings, we expected that muguet might alleviate the stress typically experienced by subjects engaged in vigilance tasks, and thereby indirectly facilitate performance accuracy by reducing the distracting effects of negative mood.

Forty-eight subjects, 24 males and 24 females, were paid for their participation and were individually tested in a 40-minute vigil.

Their task was to detect the occasional occurrence (p=.02) of visual signals appearing on a video display terminal. The signals were randomly embedded in a sequence of non-signal events, with the rate of stimulus presentation set at 24 events per minute. Non-signals consisted of a pattern of two vertical lines, each 10 mm from a centering dot. Critical signals for detection consisted of the same display, except that each dot was 12mm distant from the dot. Subjects were assigned randomly to one of three conditions: peppermint, muguet, or unscented air. The fragrances, or air, were delivered to the subjects through a modified oxygen mask in bursts of 30 seconds starting at 4.5 minutes into the vigil, and then once every five minutes thereafter.

In addition to measuring performance accuracy, we administered three scales to assess aspects of stress: (1) the Thackray Mood Scales (Thackray, Bailey, & Touchstone, 1977), which taps attentiveness, sleepiness, strain, boredom, and irritation; (2) the Yoshitake Symptoms of

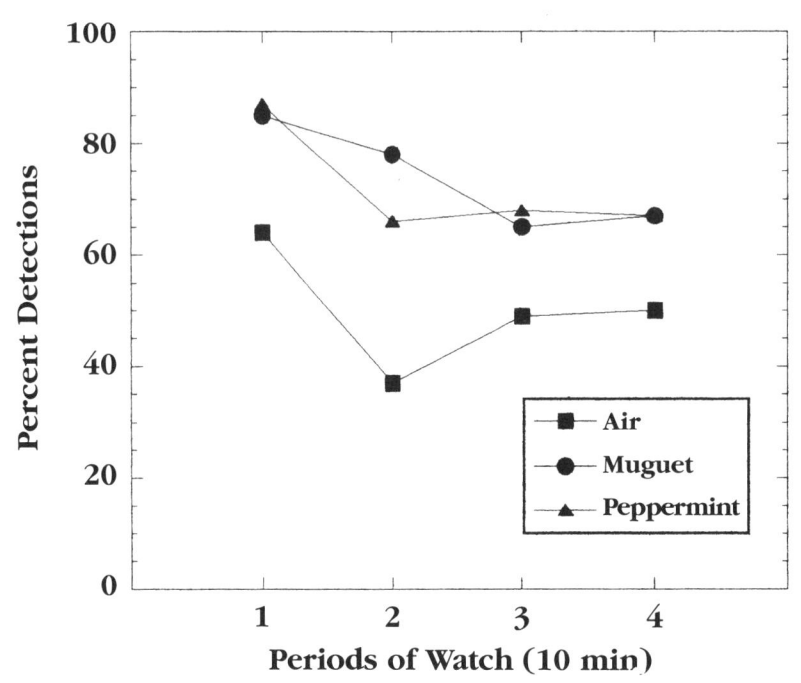

FIGURE 1

Mean percentages of correct detections for subjects in the Air, Muguet, and Peppermint conditions as a function of periods of watch.

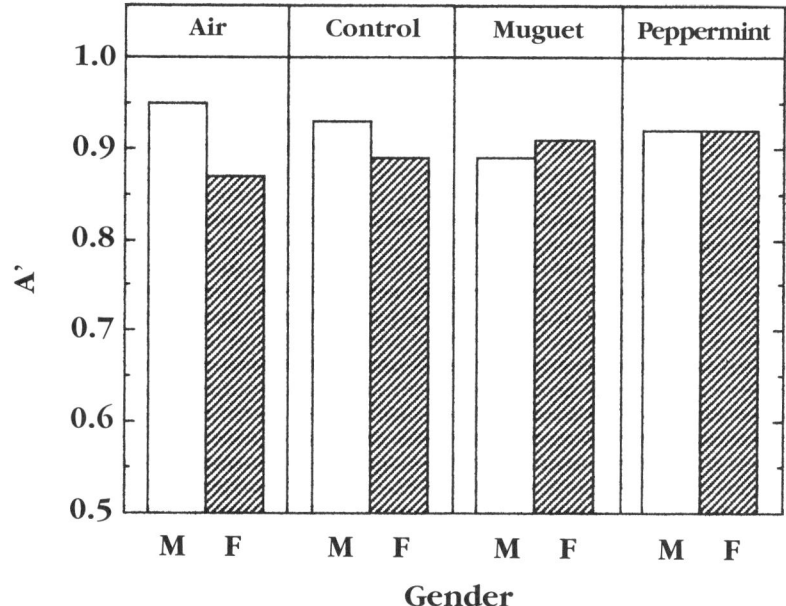

FIGURE 2

Mean perceptual sensitivity scores for males (M) and females (F) in four fragrance conditions. Note that the sensitivity index A' can range from 0.5 to 1.0.

Fatigue Scale (Yoshitake, 1978), a 30-item checklist for fatigue symptoms such as headache, dizziness, and eyestrain; and (3) the Stanford Sleepiness Scale (Hoddes et al., 1973), a Likert-type, seven-point scale for measuring alertness/sleepiness. Finally, we also administered the NASA TLX, an instrument for measuring the subjective workload associated with performing a task (Hart & Staveland, 1988). Those four instruments were given both before and after the vigil, so that pre- to post-vigil change scores could be calculated.

In essence, the results (Figure 1) showed that subjects' performance accuracy (percentage of signals correctly detected, or "hits") was significantly higher in the two fragrance conditions than in the unscented air condition.

That is, both peppermint and muguet appeared to have a beneficial effect on performance; further analysis indicated that this effect was on perceptual sensitivity rather than on subjects' willingness to report seeing the very infrequently-presented critical signals. As in other studies (cf. Warm, 1993), subjects reported that the vigilance task was stressful and demanding. However, none of the four subjective report measures revealed a significant effect of either peppermint or muguet. Thus, both fragrances helped subjects to perform more accurately, but neither made them feel better than control subjects stimulated by unscented air.

2. Effects of self-administered fragrances on performance and stress in a visual vigilance task.

The results of the first experiment were quite encouraging, especially in light of the fact that in designing it we had no a priori theoretical or empirical basis on which to select the parameters of fragrance administration. We had wanted to deliver sufficient amounts of fragrance to be effective, but not so much that subjects either adapted to the fragrance or found it aversive. In addition, for purposes of assuring that all subjects were treated alike, the fragrances and unscented air were delivered on a fixed schedule. Some subjects might thereby have received more or less fragrance than was optimal for them. We thought that we might obtain more dramatic results if subjects were permitted to deliver fragrance to themselves ad libitum, as would likely be the case in many work settings. Hence, we replicated the first experiment with the exception that subjects had control over fragrance delivery. Whenever they wished to receive fragrance, or unscented air in the control condition, they simply pressed a button which initiated an 8 sec. burst of air flow through their oxygen mask.

Three other changes were made in the procedure: (1) Subjects, again 24 men and 24 women, were not paid, but rather participated to fulfill a course requirement; (2) the probability of signal occurrence was raised from .02 to .04; (3) a second control group which did not wear the oxygen mask and received no auxiliary stimulation was added to the experimental design.

Analysis of the performance accuracy data, reported here as A, a signal detection theory index of perceptual sensitivity, revealed a more complex finding than that of the first experiment (Figure 2). Peppermint and muguet again enhanced performance accuracy, but only for the female subjects.

Women's accuracy was lower than that of the men in both control conditions, a result which is not uncommon in vigilance tasks, requiring spatial discrimination (Dittmar, Warm, Dember, & Ricks, in press). What the fragrances seemed to do was to bring the women's performance level up to that of the men, since there were no significant differences between the sexes in the fragrance conditions. Examination of the data with regard to how often the subjects pressed the fragrance-delivery button did not reveal a sex difference. And, most interesting, the subjects on average received about as much total fragrance with ad libitum delivery as they did in the previous experiment, in which the delivery schedule was experimenter imposed. Finally, we again saw no evidence of a fragrance effect on either of the subjective report measures (the Stanford and Yoshitake scales). It may be that the vigilance situation is simply too stressful to be mitigated by the relatively small doses of fragrance employed in these experiments. It is also quite possible that neither peppermint nor muguet is the right fragrance for this purpose.

This experiment partially replicated the performance accuracy findings of the initial experiment in that both peppermint and muguet enhanced accuracy; however, that result applied only to the women subjects. There is reason to believe that women are generally more sensitive to olfactory stimuli than are men (Koelega & Koster, 1974); however, no sex difference or interaction with sex was evident in the first experiment. Perhaps the more compelling conclusion from this experiment is that fragrance only helps those whose performance would otherwise be subpar. We will return to this conclusion in discussing the results of the third study.

3. Attentiveness as a moderator of the effects of fragrance on vigilance performance.

In Study 2, we found that for women, but not men, performance on a visual vigilance task improved with self-administered exposure to peppermint and muguet. In that respect, sex of subject serves as what is known as a moderator variable — one that affects the relation between other variables (in this case, the relation between exposure to fragrance and accuracy of vigilance performance). The experiment to be summarized in this section uncovered another potential moderator variable, which we refer to as attentiveness. One of the instruments used, the Thackray, had a scale for assessing subjects' self-reported degree of attentiveness at the beginning and at the end of the experiment. Virtually all subjects began the experiment with high values on attentiveness, but some indicated a considerable loss in attentiveness at the end of the vigil. Accordingly, we conducted an experiment to test the hypothesis that attentiveness would

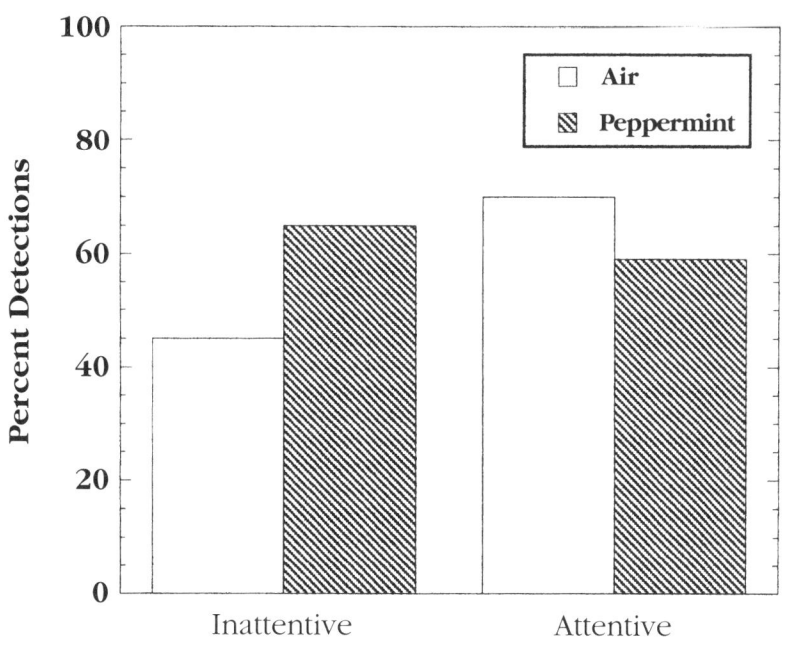

FIGURE 3

Mean percentages of correct detections in the 4th period of watch for inattentive and attentive subjects in the Air and Peppermint conditions.

moderate the effect of peppermint on performance. The results, based on 36 subjects, are displayed in Figure 3. Statistical analysis revealed that peppermint enhanced the final period performance of the 18 subjects categorized as inattentive, but not that of the 18 attentive subjects.

As was the case in Study 2 in which subjects self-administered fragrance, it seems that fragrance administration will be beneficial to performance accuracy only for those subjects who need the extra boost that fragrances provide. If in any experiment, through the vagaries of sampling, there are too many subjects in the fragrance condition who do not need that boost, the overall results will be inconclusive. Conversely, if the experimental group happens to contain a substantial number of subjects who can benefit from exposure to fragrance, the overall results will prove supportive without the necessity of taking potential moderator variables into account.

4. Brain recording, attentiveness, fragrance and performance.

The final experiment in this series pursued the issue of attentiveness further by examining the impact of peppermint on the electrical activity of the brain. Visual evoked potentials ("event related potentials": ERPs) were recorded from the scalp while subjects performed the vigilance task used in Experiment 1. Visual ERPs represent time-locked changes in voltage on the order of micro-volts that are recorded from the scalps of subjects in response to a discrete visual stimulus (in this case, the non-signal events). The potentials, as small as they are, can be extracted from the electrical noise in which they are embedded through a process of computer averaging of repeated stimuli. The ERP has a characteristic waveform, with identifiable components. Two of these, the N 160 and the P 300, have been linked to the extent to which the subject is attending to the relevant stimuli. The more attentive the subject is to the stimuli, the greater

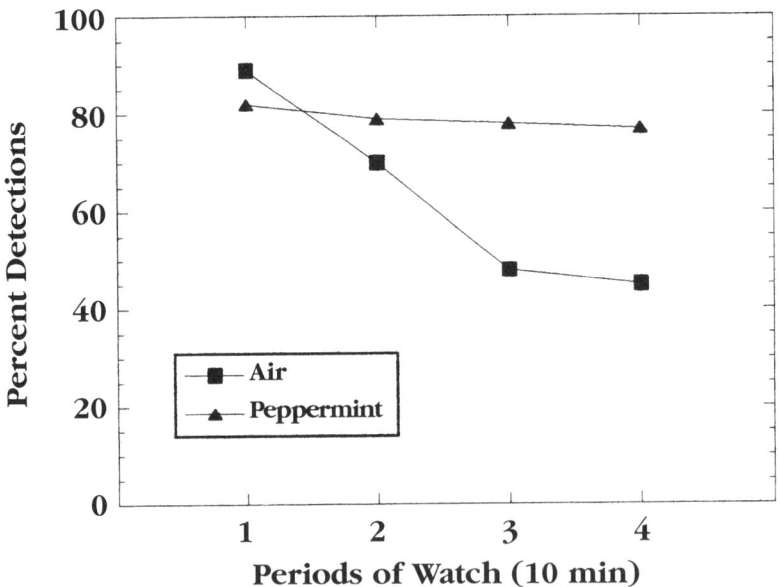

FIGURE 4

Mean percentages of correct detections for subjects in the Air and Peppermint conditions as a function of periods of watch.

the amplitude of these two wave components (Parasuraman, 1990). We report here only the data from the N 160 measurements.

In essence, the performance data confirmed the results of the first experiment (Figure 4). These results, based on 10 subjects per condition (peppermint or unscented air), were even stronger than those of the first study, in that they revealed that peppermint not only enhanced performance accuracy in general, but also eliminated the temporal decline in accuracy that is typical in vigilance research (the so-called vigilance decrement). Moreover, the ERP results (N 160 amplitude) closely matched the performance data (Figure 5). As noted earlier, these data strongly implicate an attentional process in vigilance performance, and indicate quite convincingly that the performance-enhancing effect of peppermint is linked to its ability to help subjects maintain their attentiveness, especially towards the end of the vigil. The mechanism whereby peppermint does that is yet to be determined, and we do not know whether other performance-enhancing fragrances, such as muguet, operate through the same mechanism.

In summary, the experiments reported here provide evidence that fragrances can serve to improve performance accuracy in a tedious, but demanding laboratory vigilance task. We have provided initial evidence for the possibility that this effect may be moderated by both sex of subject and self-reported attentiveness, as well as attentiveness assessed by brain recording techniques. We need to examine other fragrances, other schedules of fragrance delivery, other subject populations, and other behavioral measures, both in the laboratory and in the field.

Authors' Note

We are grateful for the assistance of Dr. Jonathan Gluckman, Paula Grubb, W. Todd Nelson, and Teresa Winhusen in carrying out these experiments.

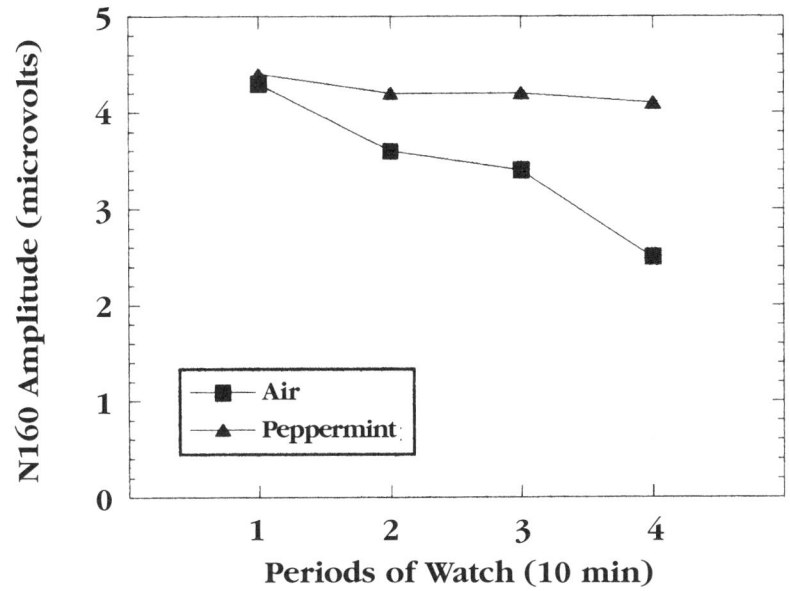

FIGURE 5

Mean N160 amplitude in the Air and Peppermint conditions as a function of periods of watch.

References

Dittmar, M.L., Warm, J.S., Dember, W.N., & Ricks, D.F. (1993). Sex differences in vigilance performance and perceived workload. Journal of General Psychology, 120, 309-22.

Hart, S.G., & Staveland, L.E. (1988). Development of NASA-TLX (Task Load Index): Results of empirical and theoretical research. In P.A. Hancock & N. Meshkati (Eds.) Human mental workload, 139-83. Amsterdam: Elsevier.

Harver, A., Katkin, E.S., Erhlichman, H., & Warrenberg, S. (1989). Autonomic and affective responses to odors. Paper presented at the meeting of the Society for Psychophysiological Research.

Hoddes, E., Zarcone, H., Smythe, R., Phillips, R, & Dement, W.C. (1973). Quantification of sleepiness: A new approach. Psychophysiology, 10, 431-6.

Koelega, H.S., & Koster, E.P. (1974). Some experiments on sex differences in odor perception. Annals of the New York Academy of Sciences, 237, 234-46.

Lorig, T.S., & Schwartz, G.E. (1988). Brain and odor: I. Alteration of human EEG by odor administration. Psychobiology, 16, 281-84.

Parasuraman, R. (1990). Event-related brain potentials and human factors research. In J.W. Rohrbaugh, R. Parasuraman & R. Johnson (Eds.), Event-related brain potentials: Basic issues and applications, 279-300. New York: Oxford.

Thackray, R.I., Bailey, J.P., & Touchstone, R.M. (1977). Physiological, subjective, and performance correlates of reported boredom and monotony while performing a standard radar control task. In R.R. Mackie (Ed.), Vigilance: Theory, operational performance, and physiological correlates, 203-15. New York: Plenum.

Warm, J.S. (1993). Vigilance and target detection. In B.M. Huey & C.D. Wickens (Eds.), Workload transition: Implications for individual and team performance, 139-79. Washington: National Academy Press.

Warm, J.S., Dember, W.N., & Parasuraman, R. (1991). Effects of olfactory stimulation on performance and stress in a visual sustained attention task. Journal of the Society of Cosmetic Chemists, 42, 199-210.

Yoshitake, H. (1978). Three characteristics of patterns of subjective fatigue symptoms. Ergonomics, 21, 231-3.

The Authors

William Dember

Dr. William N. Dember is Distinguished Research Professor of Psychology at the University of Cincinnati, Cincinnati, Ohio. He teaches courses at both the undergraduate and graduate level in motivation and emotion. His research interests are broad, including visual perception, curiosity, vigilance, optimism/ pessimism, and, of course, the behavioral and mood effects of fragrances. Dr. Dember is co-author with Joel S. Warm of The Psychology of Perception, co-author and editor of several other books and author/co-author of over 100 journal articles and book chapters.

Joel S. Warm

Dr. Joel S. Warm is Professor of Psychology and Director of the Graduate Training Program in experimental psychology at the University of Cincinnati, where he has been for the past 27 years. His research interests lie in the fields of sensation/perception and human factors/ ergonomics. He is editor of Sustained Attention in Human Performance, as well as co-author of The Psychology of Perception with Dr. William Dember.

Raja Parasuraman

Dr. Raja Parasuraman is Professor of Psychology and Director of the Cognitive Science Laboratory at the Catholic University of America in Washington D.C. Dr. Parasuraman's research interests span the areas of aging and Alzheimer's disease, attention, automation, event-related brain potentials, and olfaction. He is author/co-author of over 50 articles and publications. He is co-author of The Psychology of Vigilance and co-editor of Varieties of Attention, Current Trends in Event-Related Potential Research, Event-Related Potentials of the Brain and a special issue of Human Factors on vigilance.

Using Aroma to Reduce Distress During Magnetic Resonance Imaging

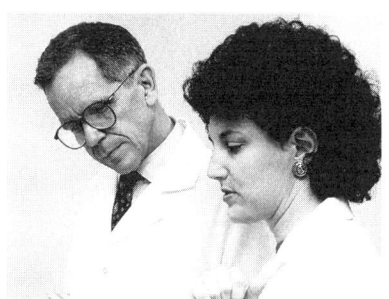

William H. Redd and Sharon Manne

William H. Redd
Sharon Manne
Memorial Sloan-Kettering Cancer Center
New York, NY

Abstract

The research investigated the use of aroma to reduce anxiety and distress in individuals undergoing magnetic resonance imaging (MRI) during diagnostic work-up for cancer. To determine the effects of this use of aroma, fifty-seven outpatients received humidified air with heliotropin (vanilla-like scent), or humified air alone through a small tube into their nostrils during the MRI procedure. Their levels of anxiety (as measured by scientifically validated patient rating scales) were compared. The delivery of heliotropin and air was contolled by a computer; the fragrance was presented in bursts to prevent adaptation. Patients who received heliotropin reported 63% less anxiety than patients who received air alone. Physiological measures (pulse and heart rate) were not affected by the presentation of the fragrance during the MRI.

The last five years has seen a rapid increase in the use of magnetic resonance imaging (MRI) scanning in the diagnosis of cancer and other medical problems. The MRI represents an important advance in medicine. It provides a method of "seeing" internal body structures, and thereby helps in the early detection of cancer and other diseases. It uses a harmless (and painless) magnetic field and radiowaves rather than x-rays. The MRI produces extremely clear images of body structures, such as the heart, brain and other vital organs. The MRI machine is large and imposing, measuring 10 feet on all sides and housed in a shielded room. During the procedure the patient is slid into a 23 inch diameter tunnel in the center of the machine and is left alone until the scan is completed. The only contact the patient has with the MRI technician is through an intercom. Depending on the problem being studied, the scan can last over one hour. In order to obtain a clear image the patient must remain still during the entire scan.

Because of the severe restriction of movement and the confined space inside the bore, some patients are likely to report feelings of claustrophobia. "It's like being in a tomb," one patient said. This fear is often heightened for individuals undergoing MRI in the diagnostic work-up for cancer. For them, the scan may confirm their worst fear: that they have cancer. When we surveyed 160 patients undergoing MRI at Memorial Sloan-Kettering Cancer Center, we found that for 10% of the patients the distress became so severe that the scan had to be terminated early (Brennan et al. 1988). Moreover, one-third of the patients reported experiencing severe anxiety while being scanned. Other studies have reported similarly high rates of anxiety, and the necessity to terminate the scan because of anxiety and distress. In one clinical study with MRI patients, Klonoff and his colleagues (1986) found that 20% of patients did not complete a scan because of claustrophobic reactions. In a study with general hospital patients, Kilborn and Labbe (1990) found that between 5 and 10% of patients became claustrophobic during the procedure.

In addition to patient distress caused by MRI, delays and early terminations of MRI scans add to the already high cost of the procedure. At a cost of $1,500 per scan, the average 15-minute delay that we found in cancer patients (Brennan et al., 1988) would result in a loss of $62.5 million nationwide each year. There is, without question, a need for cost effective methods to reduce patients' anxiety reactions during MRI scans. One possible strategy is the use of anti-anxiety medications. While they might be effective, many cancer patients want to avoid taking additonal medications, and some are especially resistant to taking drugs that affect their mental state. Beside the subjective reasons that make anti-anxiety medications less than ideal for many patients, their use can delay scans because the drugs need time to take effect. They also cause sedation which the patient must tolerate after the scan is over. Thus, while anti-anxiety medications have been used with some individuals during MRI, their use has been quite limited and is not likely to be widely adopted to reduce distress associated with MRI scans. Another strategy for making MRI scans easier is to provide music. Music has been used in various medical settings to help reduce stress, and some patients have reported that it helps. But no scientific evidence

regarding its effectiveness has been reported in the professional literature.

When we first considered the problem of how to reduce patient anxiety during MRI, we considered the practical limitations of the clinical setting and the need for method(s) that would be immediately effective and easy to implement. We thought of how our prior research on the use of behavioral relaxation and distraction to control chemotherapy side effects and pain in cancer patients might be relevant. In that work (Redd, in press; Burish & Redd, in press) we found behavioral relaxation and attentional distraction to be effective in controlling nausea and vomiting in patients undergoing chemotherapy. Patients who were trained in self-induced relaxation could effectively block the occurrence of aversive side effects. They could go into a "trance state" of deep relaxation and immediately terminate the feelings of nausea and anxiety. Our relaxation techniques are very similar to meditation and self-hypnosis. With young children we used a different strategy. Our goal was to distract the pediatric cancer patients' attention from their aversive symptoms in hopes of reducing them. The approach is simple and resembles the idea of "getting one's mind off the problem." Attentional distraction is critical in many of the techniques used to control pain and anxiety (such as hypnosis). We found that pediatric cancer patients were far less anxious and had less nausea when they were given video games to play as they waited for their chemotherapy treatment.

Unfortunately, we could not transfer either of our methods directly to the problem of MRI anxiety. First, the adult patients who were able to use relaxation to control their chemotherapy nausea had to be trained in relaxation before they were able to control their nausea in the context of the chemotherapy clinic. Such training would be impractical for MRI patients who typically do not undergo MRI scans on a repeated basis. Second, the video game procedure would be impossible within the MRI chamber. Third, both approaches are labor intensive and ultimately infeasible.

Fortunately, Lorig and Schwartz's research (1987 (a and b)) on the effects of fragrance on anxiety provided a suggestion of what to do. They observed reliable changes in physiological measures of arousal and anxiety when college undergraduates were presented with particular fragrances. Certain fragrances made the students less anxious. Although the "ideal" relaxing fragrance has not been identified, Lorig and Schwartz (1987b) found that presentation of a spiced apple fragrance (i.e., the smell of apple pie) produced electroencephalic (EEG), physiologic, and subjective evidence of relaxation in the students under conditions of quiet contemplation. Although the exact mechanism underlying their effects is not understood, at least four possibilities can be identified. First, certain fragrances may have certain inherent biological effects which elicit relaxation. Second, because of past experience particular fragrances may be differentially associated with relaxation, and thereby elicit what might be called "conditioned" relaxation (as with the "warm and relaxed" images associated with the smell of warm apple pie or wood burning in a fireplace). Such associated relaxation is understood in terms of Pavlovian classical conditioning (e.g., the dog who salivates when he hears the bell that in the past had been rung just before meat powder was delivered into his mouth). The fragrance elicits a conditioned relaxation response. Third, the fragrance may have served to distract the subjects from their worries and thereby resulted in anxiety reduction. And fourth, all three of the above may have been operating in combination.

Based on the research of Lorig and Schwartz, we reasoned that similar fragrance-induced relaxation effects might be achieved during MRI scans. We also reasoned that, in addition to the potential physiological effects of fragrance as an elicitor of relaxation, fragrance might serve to distract the subjects' attention from anxiety-eliciting stimuli. Our aim was quite simple: to examine the use of fragrance to reduce distress. We chose to work with individuals being screened for cancer because of their need for assistance, and the fact that such a study was feasible at our center. The specific goal was to determine if giving patients the fragrance heliotropin during the scan would help them relax and make the scan easier. Our choice of heliotropin was based on our survey of adults to determine their preference and reaction to various scents.

Research Procedure

The research procedure used in this study is called a randomized clinical trial: individuals are randomly assigned to a group receiving standard treatment or to a group receiving the treatment method under study. In our study, participants were individuals coming to Memorial Sloan-Kettering for MRI in their diagnostic work-up for cancer. We recruited individuals who were waiting for the MRI scan. We first explained the study, and asked them to sign a written statement of informed consent. They were then randomly assigned to the control (air alone) or experimental (aroma) group. If they did not wish to participate, they received the usual treatment.

Fifty-seven individuals participated. Eight criteria were used to select study participants. They had to: 1) be 18 to 65 years of age 2) have no prior or current chemotherapy 3) have no prior psychiatric history 4) be scheduled for outpatient treatment 5) have no history of allergic reactions to perfumes 6) have used no anti-anxiety or pain medication for the past 48 hours 7) experienced anxiety prior to the onset of the scan and 8) be able to detect the fragrance when it was presented.

Although we screened all individuals who received MRI as outpatients at Memorial Sloan-Kettering, some patients were excluded from participation for the following additional reasons: 1) they were allergic to the fragrance or refused to have a plastic tube inserted into their nose, 2) they already had their own way of controlling anxiety, or personally did not feel that they needed any help 3) they felt too nervous or nauseated to participate and wanted to be left alone or 4) they did not like the smell of the fragrance or thought that it would not help. The group of 57 included 26 men and 31 women and was representative of the patients seen at the Memorial Sloan-Kettering. Our sample was representative of all races and socioeconomic groups.

The fragrance was 5% synthetic heliotropin suspended in a 25% solution of odorless dipropylene glycol and odorless diethylpthalate. The odorant has a faint, sweet, vanilla-like scent. Heliotropin was selected after pilot testing of five fragrances supplied by International Flavors and Fragrances, Inc. (New York, New York). Among the 25 medical staff at the hospital in which the study was conducted, heliotropin was rated as the most relaxing, most pleasant and moderately intense fragrance.

Before the scan began the patients were unaware of the group to which they were assigned (control or experimental). During the scan, a plastic tube was placed into the patient's nostril through which stimuli (fragrance or air) were delivered. Delivery was controlled by a computerized valve so that the fragrance was presented in punctate bursts followed by non-fragranced bursts. The same schedule of delivery was followed for the control group, except only non-fragranced air was used. This procedure was designed to reduce the possibility of adaptation to fragrance.

Results and Discussion

We examined the effects of fragrance administration on several ratings of anxiety and physiologic functioning (heart rate and blood pressure) before and after the scan. In addition, behavioral measures (whether the patient requested termination of the scan before it was completed, and whether the scan was terminated prematurely) were obtained. Patients who were administered the heliotropin scent during MRI scans, and was rated the fragrance as pleasant, reported a significant decrease in average anxiety, while patients who received humidified air alone did not. The fragrance intervention was successful for the 70% of those patients who experienced heliotropin as pleasant.

There appear to be at least two factors that might account for the effects obtained in the present study. As we mentioned earlier, one possibility is cognitive distraction. Patients may have been distracted from feelings of anxiety during the scan by their associative reactions to the fragrance. The fragrance may have elicited reactions (i.e., thoughts, images, emotions, etc.) that were incompatible with sensations of anxiety. Prior research has shown that odor can help elicit moods and memories (Ehrlichman & Halper, 1988; Kraut, 1982; Stellar & Stellar, 1988). Thus, feelings of anxiety may have been blocked by the patient's reactions to the fragrance. Unfortunately, patients were not queried regarding their associations to the fragrance beyond whether or not they liked it. The distraction explanation is consistent with our prior research on the use of cognitive (i.e., visual) imagery to block anxiety and anticipatory nausea in chemotherapy patients (Redd, Andresen & Minagawa, 1982 cf., Redd et al., 1987). In that research, patients did not experience anxiety or nausea as long as they were engaged in the distracting task. As soon as they stopped engaging in the imagery, symptoms reappeared. The same processes may have been operating in this study.

The second possible mechanism is physiological relaxation in direct response to the physical properties of the fragrance. This hypothesis is consistent with the study by Lorig and Schwartz (1987, a & b). They found reductions in both physiological and psychological measures of anxiety in college volunteers exposed to apple-spice fragrance in a non-stress laboratory setting. However, in the present study we failed to obtain physiological changes in response to the fragrance. This failure to replicate may have been because we used heliotropin rather than apple-spice, or because we were unable (for technical reasons) to measure physiological reactions until after the scan was finished. In the Lorig and Schwartz laboratory studies, measures were obtained at the same time that the fragrance was being administered. Unfortunately, the magnetic field of the MRI chamber made such simultaneous physiological measurement impossible.

An interesting finding was the importance of the patients' subjective liking (i.e., hedonic response) to the fragrance. The benefits of fragrance administration were only apparent for the patients who found it pleasant. There are a number of possible explanations for this finding. The strength of any association (thoughts, images, emotions) may have been stronger when the fragrance was pleasant for the patient. Alternatively, patients may have used "pleasant" to mean calming and relaxing. If so, then we would predict that they would report less anxiety for the period that they were exposed to heliotropin. Another possibility is that for patients who did not rate heliotropin as pleasant, any associations (images) may have actually been negative and even served to increase anxiety.

The third set of issues to be addressed concerns the clinical use of fragrance to control anxiety during stressful medical procedures. First, were the reductions in MRI anxiety clinically important? Our data suggest a clinically significant effect: patients who received fragrance showed a 47% reduction in average anxiety whereas patients in the control condition showed a 5% reduction. Second, how might the positive impact of fragrance administration be enhanced? There are at least two approaches to improving the effectiveness of the intervention. The first would be to increase the likelihood that patients would find the fragrance pleasant and thereby show greater benefit. This could be accomplished by giving patients a choice of fragrance to be used during the scan. A second approach would be to combine fragrance administration with other anxiety reduction techniques (e.g., audiotape relaxation exercise) during the scan. Another issue is the clinical feasibility of fragrance administration during MRI scans and other stressful medical procedures. The present procedure was quite elaborate, including a computerized delivery system, separate lines for fragrance and air alone, etc. However, in a non-experimental setting where fragrance contamination and counterbalancing were critical, an easier (and less costly) system could be used. The possibility of much simpler methods of fragrance administration means that fragrance could be incorporated into routine clinical practice without great difficulty or significant expense.

Although the results must be replicated before we can recommend the routine clinical use of fragrances, the results suggest that olfactory cues may be useful in controlling anxiety and distress during stressful medical procedures.

References

Brennan, S.C., Redd, W.H. & Jacobsen, P.B., et al. (1988). Anxiety and panic during magnetic resonance scans. Lancet ii, 512.

Burish, T.G. & Redd, W.H. (in press). Symptom control in psychosocial oncology. Cancer .

Ehrlichman, H. & Halpern, J. (1988). Affect and Memory: Effects of pleasant and unpleasant odors on retrieval of happy and unhappy memories. J Pers Soc Psych, 55, 769-79.

Kilborn, C.I. & Labbe, E.E. (1990). Magnetic resonance imaging scanning procedures: Development of phobic response during scan and at one-month follow-up. J Beh Med, 13(4), 391-401.

Klonoff, E.A., Janata, J.W. & Kaufman, B. (1986). The use of systematic desensitization procedure to overcome resistance to magnetic resonance imaging (MRI) scanning. J Beh Ther Exp Psych, 17, 189-92.

Kraut, R. (1982). Social presence, facial feedback, and emotion. J Pers Soc Psych, 42, 853-63.

Lorig, T.S. & Schwartz, G.E. (1987a). Factor analysis of EEG frequencies and self-report. Psychophys, 24, 599.

Lorig, T.S. & Schwartz, G.E. (1987b). EEG during relaxation and food imagery. Psychophys, 24, 599.

Redd, W.H., Andresen, G.V. & Minagawa, R. (1982). Hypnotic control of anticipatory nausea in patients undergoing cancer chemotherapy. J Consult Clin Psych, 50, 14-9.

Redd, W.H., Jacobsen, P.B., Die-Trill, M., et al. (1987). Cognitive/attentional distraction in the control of conditioned nausea in pediatric cancer patients receiving chemotherapy. J Consult Clin Psych., 55, 391-5.

Redd, W.H. (In press). Advances in psychosocial oncology in pediatrics. Cancer.

Stellar, J. & Stellar, E. (1988). The neurobiology of motivation and reward. New York: Springer Verlag.

Acknowledgements

This research was supported by a grant from the Olfactory Research Fund, Ltd. We would like to thank Ismini Georgiades, Martin Kleber, and Jennifer Keates for collection of study data, Gerald Toohey and the staff of General Valve Corporation for assistance in development of the fragrance delivery system, Steve Warrenberg of International Flavors and Fragrances for supplying the fragrance, and the staff of the MRI department at Memorial Sloan-Kettering Cancer Center (particularly Nearlene Dillon) for their cooperation in conducting the study. We would also like to thank the patients for their participation in this study.

The Authors

William Redd

During Dr. William Redd's career he has held faculty appointments at Harvard Medical School, the Massachusetts General Hospital and the University of Illinois. Since 1994 he has been on the staff at Sloan-Kettering, where he is now a member and attending psychologist. He is also a professor at Cornell University Medical College. His primary interest is the use of behavioral procedures to control aversive symptoms of cancer and its treatment. Dr. Redd's research on the use of fragrance to control anxiety stems from his prior work on pain and symptom control which sought to explore the use of patient-initiated behavioral methods to control pain and anxiety during invasive medical procedures. Dr. Redd's research has been reported in two educational films and in numerous professional publications.

Sharon Manne

Dr. Sharon Manne is an Assistant Attending Psychologist at Memorial Sloan-Kettering Cancer Center and an Assistant Professor at Cornell University Medical School in New York City. Dr. Manne is a health psychologist who specializes in psychological issues confronted by cancer patients and their families. She has received research grants to investigate the effectiveness of psychological interventions to reduce distress during invasive medical procedures, how social support influences the psychological adjustment of cancer patients, how cancer effects the marital relationship, and how treatment compliance problems develop among children with cancer and their parents.

Olfactory Conditioning

The power of smell to evoke distant, emotion-laden memories is much remarked upon, and early studies found that odor memory might be particularly durable. These observations suggest that odor memory, in the form of learning and conditioning, might be strong enough to alter behavior. Gisela Epple, Donna Antonucci and Janet Hawley-Moore report on a study in progress with autistic children, where the goal is to use fragrance as a means of associating feelings of protection and security with novel (and therefore stressful) surroundings.

Another approach to olfactory learning was taken by Anne Schell, Ksenija Marinkovic, Michael Dawson and Shannon Davidson, who used skin conductance (a measure of physiological response) to explore classical conditioning to odor stimuli. They found that learned responses to odor generalize over time, something that does not happen with visual stimuli, but that is consistent with smells functioning as situational cues.

Do Odor Memories Influence Behavior in Children?

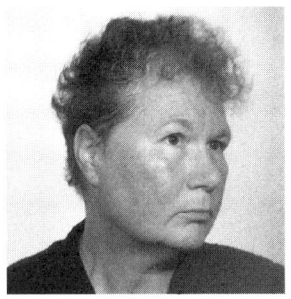

Gisela Epple

Gisela Epple
Donna Antonucci
Janet Hawley-Moore

Many readers of these lines will – at some time in their lives – have experienced vivid memories which were triggered by an odor. These memories may have been accompanied by intense emotions, emotions felt when the odor was first experienced many years ago. It has often been suggested in popular and scientific literature that our sense of smell is intimately linked to emotion, and that odors, more than any other sensory stimuli, have the ability to recall the distant past, transporting us back to situations and feelings experienced long ago. The link between olfaction and emotions may be related to the fact that the olfactory system projects directly to the limbic part of the brain, an area involved in the control of emotional behavior (Ehrlichman & Bastone, 1992; Van Toller, 1988).

Generally, it has been very difficult to investigate this phenomenon in the laboratory and there are only a handful of experimental studies reported in the literature (see Ehrlichman & Bastone, 1988). The formation of odor memories associated with emotion may well require real life experiences. These are hard to simulate in a laboratory setting. We are trying to create such experiences in our study of associations between odors and emotions in pre-school children. These studies combine the expertise of a behaviorist, a developmental pediatrician and a special education teacher. Our approach is based on the assumption that children will remember a fragrance which they experience under emotionally rewarding conditions. When this fragrance is encountered at a later time, it should elicit memories of the situation and the feelings associated with it. These memories, in turn, may influence behavior.

Our studies are conducted with autistic children, because some of these children seem to be particularly interested in the olfactory component of sensory stimuli present in their environment. Childhood autism is a developmental disorder characterized by a number of behavioral symptoms, among them severe impairment in the ability to interact with others, the inability to understand language, and the inability to communicate with speech (Wolff, 1991). If odor experiences can modify behavior, fragrances could be employed to direct the behavior of children who have difficulties in communicating verbally. Many autistic children experience extreme anxiety when routines are changed, or when they are required to go to unfamiliar places (Wolff, 1991). Fragrances that the children have learned to associate with the feeling of protection and security may reduce the anxiety they experience under stressful conditions.

In order to investigate this possibility, autistic children and their families are provided with long-lasting, pleasant fragrances in the form of customized air fresheners. The fragrance is placed in a preferred area of the child's home. The child is exposed to the fragrance for a period of 5 months, during which he or she should learn to associate the odor with the warmth and security of home.

After 5 months of exposure to the fragrance, the child's responses to a stressful, unfamiliar environment are evaluated under three conditions: (a) in the presence of the home fragrance, (b) in the presence of a different (control) fragrance, and (c) in the absence of odor. Each child visits the laboratory three times. On each occasion, the child is introduced to an unfamiliar scented or unscented test room, and stays there for 25 minutes while his or her behavior is recorded on video tape from behind a one-way mirror. The video tapes are evaluated with the help of a special software program. This provides a record of each child's behavior under the three fragrance conditions.

This study is very time-consuming and we are not yet at the point where we can predict results. The first group of children studied showed pronounced individual differences in behavior. Responses to the test situation ranged from apathy to high anxiety. These idiosyncracies are related in part to differences in the behavioral symptoms of autism. Subjects most likely to respond positively to the experimental scent are young, nonverbal children, who experience anxiety and arousal in unfamiliar environments. For these children, fragrances to which they have become conditioned under emotionally rewarding circumstances might be usable routinely to alleviate anxiety in unfamiliar places.

References

Ehrlichman, H. & Bastone, L. (1992). Olfaction and emotion. In M.J. Serby and K.L. Chobor, (Eds.), Science of Olfaction, 410-38. New York: Springer Verlag.

Van Toller, S. (1988). Odors, emotion and psychophysiology. International Journal of Cosmetic Science, 10, 171-97.

Wolff, S. (1991). Childhood autism: Its diagnosis, nature and treatment. Archives of Disease Childhood, 66. 737-41.

Acknowledgements

Fragrances were made available by Givaudan-Roure; Airwick-style disks used to deliver time-released fragrances were donated by Reckitt & Colman, Inc. The support of both companies is gratefully acknowledged. We are also grateful for additional research support from the Albert Einstein Medical Center to Donna Antonucci.

The Author
Gisela Epple

Educated in Germany, Dr. Gisela Epple is currently Member of the Monell Chemical Senses Center in Philadelphia, PA. Dr. Epple has a long term interest in the role of body scents in the control of sexual and social behavior of primates and in the regulation of female fertility and is the author of over 60 publications in this area. More recently, her research has focused on the influence of fragrances on human behavior, and on the role of carnivore scents in predation avoidance of small mammals. Dr. Epple is member of several professional organizations and Fellow of the American Association for the Advancement of Science.

SCR Conditioning with Odor and Shape CSs

Anne M. Schell

Anne M. Schell, Occidental College
Ksenija Marinkovic, University of California at Los Angeles
Michael E. Dawson, University of Southern California
Shannon Davidson, Occidental College

Introduction

The premise of equipotentiality, implied in most of the early studies of learning, assumes that the laws of classical conditioning are essentially identical for a variety of conditioned stimuli, reinforcements, and response systems, and that regardless of the sensory modality of the Conditioned Stimulus (CS) and the Unconditioned Stimulus (UCS) or the nature of the Conditioned Response (CR), the same basic rules of conditioning apply (Ohman, Fredrikson, Hugdahl, Rimmo, 1976). However, from research on "belongingness" between the CS and UCS (Seligman, 1970) and taste aversions (Garcia & Koelling, 1966), it has become clear that simple generalization of the laws of conditioning from one type of CS-UCS pairing to another may be unwarranted. This study was designed to explore whether some of the common findings regarding conditioning obtained with visual and auditory stimuli can be generalized to olfactory stimuli as well.

Olfaction is often considered "primitive and elemental," due to its evolutionary primacy, because phylogenetically the sense of smell was developed before either vision or hearing (Stoddart, 1982). The unique role of olfaction is apparent anatomically and physiologically as well as behaviorally. Unlike most sensory cells, the cells of the olfactory epithelium perform both primary reception and conduction functions. That is, they transduce chemical stimulation into neural impulses and at the same time they carry the impulses to the olfactory bulbs via their axons, which form the olfactory nerve. Olfactory central connections also differ from those of other sensory inputs, since projection is directly to the olfactory cortex rather than through the common thalamo-neocortical relay. These direct connections between the receptor and the olfactory cortex are unique in that they place the hypothalamus at a distance of just two synapses from the peripheral olfactory input. Other primary sensory systems can only influence the hypothalamus through a far more indirect, hippocampal pathway (Scott & Pfaffman, 1967). Direct connections with the hypothalamus are important because of this structure's control of the autonomic functions of the organism, with direct ramifications for emotional and motivational behavior (Grossman, 1973).

Evidence reviewed by Cain (1974) indicates that the olfactory bulb also interacts with limbic system structures that mediate motivated and emotional behaviors. Close anatomical connections between the olfactory system and limbic structures may be the substrate of the "emotionality" of the olfactory system. Thus, evidence indicates that olfaction as a sensory modality has several unique aspects, and therefore learning parameters based on auditory and visual stimuli may not be valid for the olfactory system. In particular, because of the more direct connections between the olfactory system and hypothalamic and limbic areas, compared to the connections between these centers and other sensory systems, the powerful influences of cognitive processes which have been found to characterize human autonomic classical conditioning with visual and auditory stimuli may not characterize conditioning with olfactory stimuli.

For instance, with visual or auditory stimuli it has generally been found that in normal human subjects classical conditioning of autonomic responses does not occur unless subjects are aware of the contingency between the conditioned stimulus (CS) and the unconditioned stimulus (UCS), that is, unless they are able to recognize or verbalize the CS-UCS relationship (see Dawson & Schell, 1982, for a review of this literature). Moreover, although conditioned autonomic responses are well-retained over time, laboratory investigations have indicated that they are retained only by those subjects who retain their memory of CS-UCS relationships; those subjects who forget which CSs predict USCs do not retain their conditioned responses (Hammond, Baer & Fuhrer, 1980). Finally, there is evidence that when a time interval intervenes between original acquisition of a conditioned response and presentation of extinction trials, the conditioned response does not persist past the point of cognitive extinction, that is, past the point where the subject no longer expects the UCS to occur after the CS, at least unless the conditioned stimulus is of a "biologically prepared" or "potentially phobic" nature (Schell, Dawson, & Marinkovic, 1991).

Classical conditioning using olfactory stimuli as CSs has been little investigated with human subjects. However, Marinkovic, Schell, & Dawson (1989) carried out classical conditioning of the skin conductance response using odor stimuli and a CS-UCS contingency masking task and demonstrated that as with auditory and visual stimuli, awareness of the CS-UCS contingency was necessary in order for conditioning to occur, and differential conditioning was observed only at and after the point at which contingency awareness occurred. The present study was designed to determine whether the strength of other cognitive variables at several stages of the conditioning and extinction process was as strong for olfactory conditioning as for conditioning with other sensory modalities.

Experimental Questions

(1) There are anecdotal reports that responses conditioned to olfactory stimuli tend to be unusually persistent over time. Can it be demonstrated in a laboratory setting that retention of the SCR CR over a period of time is better with olfactory CSs than with visual CSs?

(2) Will the CR conditioned to olfactory CSs be retained by subjects who have forgotten the CS+ /UCS contingency?

(3) Will the CR conditioned to olfactory CSs persist after the point of expectancy extinction?

Methods

Subjects

Subjects were 47 volunteers, aged 18-22, 32 women and 15 men.

Design

All subjects were presented with a classical discrimination conditioning paradigm with Adaptation and Acquisition trials during Session 1, and Retention, Reconditioning, and Extinction trials during Session 2. Each subject was conditioned using both a set of odor and a set of visual CSs, with one of four odors and one of four shapes being followed by the shock UCS.

Stimulus Materials and Apparatus

Visual CSs consisted of slides of random shapes. These computer-generated shapes are smooth-contoured abstract forms of a rather ameboid character. They were projected by a carousel projector onto a screen in front of the subject. The odor stimuli used were methyl salicylate (wintergreen), eugenol (clove oil), rose oil, and lavender oil. The odorants were diluted for presentation in 10 ml of mineral oil and were presented using a device developed in the Occidental College Psychophysiology Laboratory.

Four vials containing odors and one containing odorless mineral oil were enclosed in a box placed on a tray in front of the subject's nose. The box housed five solenoids which were energized by an IBM AT computer-controlled external power source. A clip attached to the end of each solenoid held a 6 cm glass vial with an odorant. The solenoids were mounted on the arc of a circle, so that when one was energized, its vial was moved forward into position under a hole in the top cover of the box. The box was positioned so that the hole in the top cover was just below the subject's nose. A red light mounted in front of the subject indicated the presentation of an odor and signaled the subject to inhale and to attend to the odor.

The DC electric shock UCSs were delivered from a Grass S-9 stimulator and administered to the subject's left leg through silver electrodes. Skin conductance was recorded on a Narco-Bio DMP 4 B Physiograph through a constant .5 V bridge. SCRs were recorded from Beckman silver-silver chloride electrodes filled with isotonic electrode paste placed on the volar surface of the first and third fingers of the left hand. Stimulus onsets and durations for the slides and odors (5.0s), the shock UCS (0.5s), and the light which signalled trial onsets were controlled by the lab computer. Throughout the experiment, subjects indicated expectancy of the UCS during each CS presentation using a series of buttons which were labelled to represent a seven-point scale ranging from absolutely certain that no shock was about to come through to absolutely certain that shock was about to come.

Procedure

Session 1: Subjects were told that the purpose of the experiment was to measure their physiological responses to different visual and olfactory stimuli. They were told that they would see pictures of abstract shapes, smell various odors, and feel an occasional electric shock to their leg, which they would set individually at a level of "strongly annoying, but not painful." They were further told that the shock would usually, but not always, follow a particular one of the shapes and a particular one of the odors. They were asked to be sure to notice which stimuli were associated with the shock.

Each trial consisted of both a slide and an odor presentation. Vials with detectable odors were paired with blank grey slides, and slides with shapes were paired with vials containing only mineral oil. Each trial began with the onset of a red light which preceded the slide and the simultaneous movement of an odorant-containing vial into position by 1.0s. This signalled the subject to lean forward slightly to smell the odor stimulus while simultaneously watching the screen for a slide. Subjects indicated their expectancy of the UCS during each CS by using the buttons. Subjects were presented with sixteen Adaptation trials (two presentations each of each SC) and 80 Acquisition trials (ten presentations each of each CS). The odor and shape CS+s were reinforced on seven of their ten presentations. For each CS+ another CS in the same modality was designated as CS-.

Session 2: The second session was scheduled three months after the first. Subjects were first presented with 16 Retention trials of the same type as during Session 1 during which each of the eight CSs was presented trice. Subjects used the button box to indicate their memory of whether or not each CS had been followed by the UCS during Session 1. Following the Retention trials, subjects received 20 presentations each of the odor and shape CS+ and CS-. On its first presentation, each odor and shape CS+ was followed by the UCS (Reconditioning). On the following 19 Extinction presentations, no further shocks were delivered.

Results

The results reported below are for the FIR SCR, the response beginning 1.0 - 4.0 s after CS onset.

Adaptation

A Sex x CS Modality x Conditioning (CS₁ versus CS-) x Trial Block ANOVA of the Adaptation SCR revealed a significant Modality effect, $F(1, 45) = 14.15$, $p < .001$, with responses to odor CSs being larger than those to shape CSs. However, there was no even marginally significant tendency for subjects to initially respond more to CS+ than to CS-.

Acquisition

Button Expectancy Data: Data were analyzed only for subjects who were able to correctly identify both the odor and the shape CS+s. Subjects were classified as aware or unaware of a contingency based on the trial-by-trial button expectancy data, being categorized as aware of the contingency if they expressed both a positive expectancy of the UCS during CS+ and a negative expectancy of the UCS during CS- for three consecutive pairs of trials.

A Sex x Modality x Conditioning x Blocks (of 2 trials) ANOVA was carried out for the button expectancy data of the aware subjects. Expectancies associated with CS+ and CS- in both modalities averaged across trial blocks are shown in the left panel of Figure 1. The Conditioning effect was highly significant, $F(1,45) = 1177.69$, $p < .001$. The Modality x Conditioning interaction was also significant, $F(1,45) = 12.41$, $p < .001$; differential expectancies between CS+ and CS- were somewhat greater for the shape CSs than for the odor CSs, indicating slightly greater ease of discrimination between shape stimuli than odor stimuli.

A similar ANOVA carried out for the Acquisition SCR data of the aware subjects revealed a highly significant effect of Conditioning, $F(1,45) = 68.00$, $p < .001$. Acquisition SCR data are shown in the right panel of Figure 1. The Modality x Conditioning interaction was also significant, $F(1, 45) = 10.51$, $p < .003$. Discriminative responding was better with the shape than with the odor CSs, paralleling the slightly greater expectancy differential with shape than with odor stimuli. In order to establish that significant differential responding to CS+ and CS- had occurred for both modalities, t tests were carried out comparing the average responses to CS+ and CS- in both modalities. Both tests yielded highly significant results.

Retention

One experimental question of great interest concerned whether or not the differential SCR to

CS+ and CS- would be retained with these CSs when the original differential expectancy of the UCS after CS+ and not after CS- had been forgotten. Therefore, we divided the subjects who were originally aware of both contingencies into those who did and did not remember the contingency for each modality. A subject was classified as remembering a contingency if he or she indicated a higher expectancy of the UCS following CS+ than CS- during both of the Retention trial blocks, and was otherwise classified as not remembering the contingency. Of the 47 subjects, 29 remembered the shape contingency and 25 remembered the odor contingency.

To examine retention of the CR to the shape CSs, a Memory (rememberers versus forgetters) x Sex x Conditioning x Trial Block ANOVA was performed on the Retention SCR data. Average responses of rememberers and forgetters to the shape CS+ and CS-, along with their button expectancy averages, are shown in the top panel of Figure 2. The effect of Memory was highly significant, $F(1,43) = 10.81$, $p < .001$, as was Conditioning, $F(1,43) = 8.43$, $p < .006$, and the Memory x Conditioning interaction, $F(1,43) = 10.41$, $p < .003$. As is evident from Figure 2, subjects who remembered the contingency retained the differential SCR and those who forgot it did not. In order to verify specifically that retention of the differential SCR was significant among those who remembered the shape contingency and was not among those who did not, t tests were performed comparing responses to the shape CS+ and CS- in each group. The difference was significant for the rememberers but not for the forgetters.

A parallel analysis was performed to examine the effects of memory for the odor contingency. The results of this analysis were markedly different from those for responses to the shape CSs. Surprisingly, neither the Conditioning effect, the Memory effect, nor the Conditioning x Memory interaction approached significance. Average responses of rememberers and forgetters to the odor CS+ and CS-, along with their button expectancy averages, are shown in the bottom panel of Figure 2. As can be seen, subjects remembering the original odor CS+/UCS contingency show nearly identical responses to CS+ and CS-. However, odor and shape contingency rememberers discriminate CS+ and CS- to almost identical degrees using the button expectancy measure.

Examination of Figures 1 and 2 suggests that the failure of the odor rememberers to discriminate the odor CS+ from the CS- is largely due to a rise from Acquisition to Retention in the response to the CS-. In order to investigate this possibility, we calculated incubation scores for the odor and shape CS+ and CS- for rememberers of the contingencies. Incubation scores were calculated by taking the difference between the average response to that CS during Acquisition, for both the SCR and the button expectancy measure. Average SCRs of rememberers of the odor and the shape contingency during Acquisition and Retention are shown in Figure 3. As can be seen, the change in SCR to the odor CS- is larger than that to odor CS+ or to either shape CS. T tests on SCR incubation scores revealed that the increases in response to the odor CS- from Acquisition to Retention was significant, $t(24) = 2.07$, $p < .05$, while the declines in response to the odor and shape CS+s and the rise in response to the shape CS-s were not. In contrast, for the button expectancy incubation scores, only the rise in expectancy of the shock after the shape CS- was significant, $t(28) = 2.49$, $p < .02$.

Reconditioning

Following the Retention trials, subjects were given one presentation each of the odor and shape CS+ and CS-, with the CS+s being rein-forced. The next presentation of each odor and shape CS+ and CS- allowed measurement of Reconditioning. A Sex x Modality x Conditioning ANOVA of the Reconditioning SCR data revealed significant effects of Modality, $F(1, 45) = 9.50$, $p < .004$. Responses to the odor CSs were larger than those to shape CSs and responses to the CS+s were greater than to CS-s. The Conditioning x Modality interaction was not significant, and t tests indicated that responses to CS+ were significantly greater than to CS- for both odor and shape CSs. The one reinforced trial for each CS+ was thus able to re-establish significant differential responding in both modalities.

Extinction

The 18 presentations of each original odor and shape CS+ and CS- which followed the trial block on which Reconditioning was assessed were the Extinction trials. A Sex x Modality x Conditioning x Trial Block ANOVA of the Extinction SCR data yielded a significant effect of Modality, $F(1,45) = 37.73$, $p < .001$, with CRs to odor CSs continuing to be larger than those to shape CSs, and a significant Conditioning effect, $F(1,46) = 24.91$, $p < .001$. T tests comparing SCRs to CS+ and CS- were significant for both the odor and the shape CSs. The effect of Trial Block was also highly significant, as would be expected, $F(8,360) = 145.23$ $p < .001$, with SCR magnitude declining over trials. The Conditioning x Modality x Trial Block interaction was also significant, $F(8,360) = 2.78$, $p < .02$. As can be seen in Figure 4, for shape stimuli, discrimination between CS+ and CS- declines fairly steadily from early to late extinction, as would ordinarily be expected. For odor CSs, discrimination at the outset of Extinction is less than for shape CSs, but the existing discrimination is better maintained over trials. Thus, differential responding to CS+ and CS- is much greater with shape CSs than with odor CSs during the first third of Extinction, but differential responding is actually greater with odor CSs than shape CSs during the last third.

SCR Conditioning Pre- and Post-Cognitive Extinction

In order to determine whether or not SCR CRs to odor or shape CSs were retained after the point of cognitive extinction, a subset of subjects was identified for whom cognitive extinction with either modality occurred. These were subjects who showed differential expectancy of the UCS after CS+ and CS- at the outset of Extinction, but during Extinction reached the point of consistently expressing equal expectancy after CS+ and CS-. Cognitive extinction was reached by 10 subjects for the odor contingency and 12 subjects for the shape contingency.

Sex x Pre- versus Post-Cognitive Extinction x Conditioning ANOVAs revealed significant Pre- vs Post-Cognitive Extinction x Conditioning interactions for both the odor data ($F(1,8) = 16.74$, $p < .004$) and the shape data ($F(1,10) = 4.98$, $p < .05$). As can be seen in Figure 5, greater responding to CS+ than to CS- for both modalities occurred only before cognitive extinction, and not after. T tests indicated significant discrimination pre-cognitive extinction in both modalities, while discrimination post-cognitive extinction did not approach significance in either modality.

Conclusions

Unexpectedly, differential SCR CRs conditioned to odor CSs were not retained as well as those conditioned to shape CSs, either by the whole group of subjects or by those who remembered the original contingency. This was due to a marked and significant rise in the response to the odor CS- from Acquisition to Retention testing three months later, a rise which did not occur in the response to the shape CS-. Thus, discrimination of the odor CS+ and CS- did not fail because subjects failed to respond to either CS, it failed because they gave large responses to both CSs. The significant rise in response to the odor CS- may suggest an incubating generalization process in which responses conditioned to the CS+ during Acquisition generalize over time to other stimuli present during the original conditioning situation. This was not observed with the visual stimuli, nor have we observed it with visual stimuli in past work. Odor stimuli may have a greater propensity than visual stimuli to function as situational cues, acquiring CRs because they characterize environments in which strongly pleasant or unpleasant stimuli are present rather than functioning solely as close temporal indicators of the arrival of such stimuli.

Investigation of SCR CR retention past the point of cognitive extinction did not indicate such retention with either the visual or the olfactory CSs. Thus retention of CRs conditioned to odor stimuli such as these does not appear to be any less influenced by cognitive processes than is retention of CSs to neutral visual stimuli. In light of findings regarding retention of CRs using visual CSs which are "fear relevant,""bio-logically prepared," or "potentially phobic," it would be intriguing to investigate retention with similarly "fear relevant" odor CSs.

References

Cain, D.P. (1974). The role of the olfactory bulb in limbic mechanisms. Psychological Bulletin, 81, 654-71.

Dawson, M.E. & Schell, A.M. (1982). Electrodermal responses to attended and nonattended significant stimuli during dichotic listening. Journal of Experimental Psychology: Human Perception and Performance, 8, 315-24.

Garcia, J., & Koelling, R.A. (1966). Relation of cue to consequence in avoidance learning. Psychonomic Science, 4, 123-4.

Grossman, S.P. (1973). Essentials of Physiological Psychology. New York: John Wiley & Sons.

Hammond, G.S., Baer, P.E., & Fuhrer, M.J. (1980). Retention of differential autonomic conditioning and memory for conditional stimulus relations. Psychophysiology, 17, 356-62.

Marinkovic, K., Schell, A.M., & Dawson, M.E. (1989). Awareness of the CS-UCS contingency and classical conditioning of skin conductance responses with olfactory CSs. Biological Psychology, 29, 39-60.

Ohman, A., Fredrikson, M., Hugdahl, K., & Rimmo, P.A. (1976). The premise of equipotentiality in human classical conditioning: Conditioned electrodermal responses to potentially phobic stimuli. Journal of Experimental Psychology: General, 105, 313-37.

Schell, A.M., Dawson, M.E., & Marinkovic, K. (1991). Effects of potentially phobic conditioned stimuli on retention, reconditioning, and extinction of the conditioned skin conductance response. Psychophysiology, 28, 140-53.

Scott, J.W., & Pfaffman, C.P. (1967). Olfactory input to the hypothalamus: Electrophysiological evidence. Science, 158, 1592-4.

Seligman, M.E.P. (1970). On the generality of the laws of learning. Psychological Review, 77, 406-18.

Stoddart, D.M. (1982). The developing mammalian olfactory system and its role in mediating the effects of psychotropic and physiologically active odorous agents. Pharmacological Theory, 17, 65-83.

The Author
Anne M. Schell

Born and raised in Waco, Texas, Dr. Anne Schell has been a member of the faculty at Occidental College in Los Angeles, CA since 1971. Dr. Schell's area of research is psychophysiology, the study of mind-body interactions and the investigation of mental states using physiological measurements. She has done research on attention deficit disorder and schizophrenia, and on aspects of attention in normally functioning persons as well as in psychopathology. A continuing research interest has been in conscious and unconscious processes in emotional learning, which led to the development of the study reported here.

Scent and Social Behavior

IV

One of the most exciting, but difficult to study, areas of olfactory research involves the effects of scent on social behavior. The Olfactory Research Fund has sponsored studies by Susan Knasko, in which she examines how ambient fragrance alters the time people spend on activities such as viewing a museum exhibit. The degree to which the scent is congruent with the physical setting appears to be a determining factor.

One of the major reasons for wearing fragrance is to make a social statement about yourself to other people. John Nezlek and Glenn Shean explored the impact of fragrance on social interactions by means of a naturalistic study in which participants kept a diary of their social interactions. Interactions were rated more positively when a person felt that others found their fragrance appealing. Fragrance use was also related to an increased salience of non-verbal social cues.

The role of fragrance in the management of social impressions has been investigated in detail by Mark Snyder and Mark Attridge. Some fragrance users select perfume based on the social image they wish to project, while others choose perfume for their own personal enjoyment. These groups respond differently to fragrance advertising, and have different purchasing habits as well. The results support the psychological concept of "fragrance involvement".

Congruent and Incongruent Odors: Their Effect on Human Approach Behavior

Susan C. Knasko
Monell Chemical Senses Center

Approach behavior in an environment includes such behaviors as going into the setting, staying there, exploring the surroundings, and liking and interacting with the objects and people in the setting. Avoidance behavior is the opposite. Environmental stimuli are believed to bias approach-avoidance behavior mainly by influencing the emotional responses of pleasure and arousal, which in turn are said to affect behavior (Mehrabian & Russell, 1974). According to the theory, the more pleasant an environment, the greater the approach behavior that will occur in it (Mehrabian, 1980). This theory was supported in a previous study, where customers stayed longer in two sections of a store when the areas were scented with pleasant scents, compared to when the areas were not scented (Knasko, 1989).

The findings of that study raised several questions. First, can the presence of pleasant ambient odors increase lingering time and other approach behaviors in different types of settings? And second, how does the congruency or incongruency of the odor to the setting influence approach behavior? In the previous study the odors were pleasant, but did not really match or mismatch the setting. In many situations, however, it is possible to think of

odors that fit the setting very well or that do not fit, even though they are pleasant. If odors are going to be added to a setting to increase approach behavior, it is important to know whether any pleasant odor will accomplish this goal or whether the congruency of the odor to the setting plays a role (in a desirable or undesirable direction). Knowledge of the effect of congruency may be especially important in situations where an odor may be congruent with one aspect of a setting but incongruent with another (e.g., A scent strip in a magazine may be congruent with one of the advertisements but incongruent with an ad on the opposing page).

To address these issues, three studies were initiated. The first objective of these studies was to determine if the presence of pleasant ambient odors can increase lingering time and other approach behaviors in different types of settings. The second objective was to explore how the congruency or incongruency of the odor to the setting influences approach behavior.

The first two studies were conducted in laboratory settings. In the first study subjects viewed and evaluated photographic slides when the ambient odor of the testing room was congruent with some of the slides and incongruent with others. The second study tested the effects of congruent and incongruent ambient odors on consumer decision-making when the odor of the testing room matched or did not match the choices in a computer program. The third study, conducted in a field setting, was designed to test the effects of ambient odor on behavior in a noncommercial, public environment. An exhibit in a museum was scented with odors that were congruent or incongruent with the artifacts on display. What follows is an overview of these studies.

Study 1: Photographic Slides

The results of this study are reported in a manuscript which is currently under review (Knasko, submitted).

Methods

Ninety subjects, 18 to 35 years old, were randomly assigned to one of three odor conditions: no experimental odor, baby-powder odor or chocolate odor (IFF odors #3367-HS and #3372-HS, respectively). In pilot testing, both odors were rated as very pleasant and of slight to moderate intensity. In the baby powder condition the testing room was scented with eight perfume blotters (four in the chocolate condition), each containing 0.12 g of the odorant.

In the first part of the study, subjects looked at a series of 24 photographic slides, at their own pace, in random order. Unknown to the subjects, the computer recorded the amount of time they looked at each slide. Of the 24 slides, 6 were of chocolate items, 6 were of babies, and 12 were control slides (of pine trees and the Orient). In pilot testing the 6 chocolate slides had been rated as strong matches for the chocolate scent, and as strong mismatches for the baby-powder scent; the reverse was true for the baby slides. The control slides had been rated as mismatches for both the chocolate and the baby-powder scents.

In the second part of the study, subjects looked again at the 24 slides, and rated how pleasant and interesting they found each slide, and how easy it was to imagine themselves in the setting of each slide. The computer then prompted them to complete a mood questionnaire (which measured their feelings of pleasure and arousal), a health questionnaire, and an environmental-quality questionnaire.

Results

Subjects exposed to a pleasant room scent looked significantly longer at the 24 slides compared to subjects exposed to no experimental odor. Ratings of the slides were, for the most part, unaffected by the odor conditions.

In all three conditions subjects reported being in a positive mood. Those who were exposed to an odor were in a significantly more pleasant mood than subjects exposed to no odor. Arousal was affected by the type of odor present. Chocolate odor resulted in reports of greater arousal compared to no odor.

The number of symptoms reported varied among the three odor conditions. Subjects in the baby powder condition reported fewer health symptoms than subjects in the no odor condition. Subjects in the baby powder and chocolate conditions rated the room as smelling significantly more pleasant than those in the no odor condition.

Discussion

The results of this study support our previous finding that the presence of a pleasant room odor can increase lingering time. Congruency of the odors to the slides did not play a role in this study. The findings suggest that pleasant odors may have some general effects on humans due to their hedonic value, while other effects may be due to associations individuals have with particular odors.

Study 2a and 2b: Product Choice

These studies were conducted in collaboration with Dr. Deborah Mitchell at Temple University and Dr. Barbara Kahn at The Wharton School, University of Pennsylvania. The results of these studies are reported in a manuscript which is currently under review (Mitchell, Kahn & Knasko, submitted).

Methods

These two studies looked at the effect of congruent and incongruent pleasant ambient odors on a single product choice (Study 2a) and a series of product choices (Study 2b).

In both studies there were three odor conditions (no experimental odor, chocolate scent, or floral scent) which were crossed with 2 product classes (floral arrangement or chocolate assortments). College students (77 in Study 2a, 78 in Study 2b) were randomly assigned to one of the six conditions. In the control conditions subjects made decisions of flower arrangements or chocolate assortments when there was no odor in the room. In the congruent odor conditions subjects made decisions about the product class which matched the odor of the room (e.g., they chose a flower arrangement when the room was scented with floral odor). In the incongruent odor conditions subjects made decisions about a product class that did not match the odor of the room (e.g., they chose a flower arrangement when the room was scented with chocolate odor).

In the chocolate odor condition the room was scented with four blotters (each with .08 g of odorant), placed at each of 12 work stations (IFF chocolate-cookie odor #3372-HS). In the floral condition each station was scented with two odor-disks and four blotters (each with .06g of odorant; Givaudan-Roure fragrance #DL-700 and raw material #77-0809). In pilot testing both odors were rated as very pleasant and easily recognizable. The chocolate odor was rated as being very congruent with a candy store setting and very incongruent with a floral shop setting; the opposite was true for the floral scent.

In Study 2a subjects were given a computer scenario where they had to choose which of four flower arrangements (or chocolate assortments) they liked best, based on information they acquired concerning six attributes. They could examine as much or as little of the information about the choices as they desired. After making their choice, they completed a free-recall memory task and a questionnaire concerning the room odor. Measures included the pattern and amount of information search conducted.

In Study 2b subjects were given a computerized scenario in which they had to choose from a list of seven flower arrangements (or chocolate assortments) 21 times. Measures included several variety-seeking behaviors. After they made the 21 choices, subjects had to indicate the similarity between all pairs of the alternatives. The also completed a mood and an odor questionnaire.

Results

In Study 2a subjects in the congruent odor condition spent more time on the entire task and per acquisition than subjects in the incongruent odor condition. In the congruent odor condition subjects were more holistic in their processing (looking more evenly at all the attributes) and were more likely to choose the least preferred option (thus spreading their choices more evenly over the four alternatives), than subjects in the incongruent odor condition. Subjects in this group also generated more self-references in the memory task than subjects in the incongruent odor condition. Subjects in the no odor condition spent more time per acquisition and on the total task than did subjects in the odor conditions.

In Study 2b subjects in the congruent odor condition switched more among the seven choices, chose items that were rated more dissimilar, and chose the least favorite item more frequently, than did subjects in the incongruent odor condition. Subjects in the various odor conditions did not differ in their ratings of pleasure or arousal.

Discussion

The majority of findings in Studies 2a and 2b involved differences between subjects in the congruent and incongruent odor conditions; subjects exhibited more exploratory (approach) behavior in the congruent odor condition. The role of congruency in these findings may be due to these studies requiring cognitive involvement and/or subjects experiencing a room odor that was completely congruent or incongruent with their task.

Study 3: Museum Exhibit

The results of this study are reported in a manuscript that is currently in preparation (Knasko, in preparation).

Methods

An exhibit in an anthropology museum, displaying the crafts and clothing of the native Alaskan Indians, was scented four days a week (Tuesday-Friday) for eight weeks. There were four odor conditions: no room odor, incense odor, bubble gum odor and leather odor (Givaudan-Roure odors #77-2439, #TAA27062 and #TAA27061, respectively). Each day of the week was scented twice with each odor, in a block random order. In pilot testing the incense and bubble gum odors were rated as smelling very pleasant; incense was rated as being congruent with the display, while bubble gum was rated as being incongruent. Although this study focused on pleasant odors, leather was included to pilot the effects of an unpleasant but congruent odor. Scenting was accomplished through the use of odor disks placed in inconspicuous places and with odor blotters placed in electric and battery-operated fans.

A surveillance camera video-taped visitors as they walked through the exhibit. The video tapes were scored by individuals unaware of the odor condition on the day of taping. The tapes were scored for a number of factors (e.g., sex of visitor, length of time they looked at the exhibit, and whether or not they were alone). Exit interviews were conducted for four weeks after the video taping was completed, when the room continued to be scented in the above manner.

Results

A more positive mood was reported by visitors in the bubble gum condition compared to those in the leather or no odor conditions. Visitors exposed to incense odor reported that they had learned more from the exhibit than visitors exposed to no odor. Visitors in the incense condition also reported that the odor of the room had a more positive influence on their enjoyment of the exhibit compared to visitors in the other odor conditions. Odor condition interacted in complex ways with a number of variables to influence lingering time.

Discussion

In this study both odor hedonics and odor congruency influenced approach behavior. In this field setting where a number of real-life variables could influence behavior, odor also interacted with non-olfactory variables to influence some aspects of approach behavior.

References

Knasko, S.C. (1989). Ambient odor and shopping behavior. Chemical Senses, 14, 718, A94.

Knasko, S.C. (submitted). Pleasant odors and congruency: Effects on approach behavior.

Knasko, S.C. (in preparation). Lingering time in a museum in the presence of congruent and incongruent odors.

Mehrabian, A. (1980). Basic Dimensions for a General Psychological Theory.

Cambridge, MA: Oelgeschlager, Gunn and Hain, Publishers, Inc.

Mehrabian, A. and Russell, J.A. (1974). An Approach to Environmental Psychology. Cambridge, MA: MIT Press.

Mitchell, D.J., Kahn, B.E. and Knasko, S.C. (in press), There's something in the air: Effects of ambient odor on consumer decision-making. Journal of Consumer Research.

Acknowledgements

Appreciation is expressed to Connie Papazickos for running the subjects and to Dr. Michael Tordoff for setting up the computerized slide program in the first experiment. The help of Connie Papazickos, Carolyn Odell, Alexa Yi, and Deborah Wong in setting up and scoring the video-tapes in the third study is gratefully acknowledged. A special thanks to International Flavors and Fragrances (Union Beach, NJ) and Givaudan-Roure (Teaneck, NJ) for supplying odorants used in these studies.

The Author
Susan C. Knasko

Dr. Susan Knasko is a Senior Research Associate at the Monell Chemical Senses Center, a non-profit institute dedicated to basic research on smell and taste. In both laboratory and field settings, Dr. Knasko, an environmental psychologist, investigates the relationship between odor and consumer decision-making, work performance, approach/avoidance behavior, and emotional responses. She is frequently consulted by representatives of consumer groups, government agencies, corporations, scientific associations and the media concerning aspects of olfaction and the effects of odor exposure.

Fragrance Use and Social Interaction

Glenn D. Shean and John B. Nezlek

John B. Nezlek
Glenn D. Shean
Department of Psychology
College of William & Mary, Williamsburg, VA

Abstract

Two studies were conducted to examine the roles that fragrances play in naturally-occurring social interaction. In each study, participants used a social interaction diary to describe their social contacts. These descriptions included participants' reactions to interpersonal events and their perceptions of how salient and how pleasing their fragrances were to the others. The results of both studies were similar: The more pleasing people believed others perceived their fragrances to be, the more positively people felt about their social interactions. In the second study, participants also described how important non-verbal, para-verbal, and verbal cues were to them during each interaction. The results suggested that non-verbal cues are more salient in social interactions in which fragrances are present.

Introduction

For centuries people have used fragrances to enhance their social interactions. However, despite this widespread use, little is known about how fragrances influence social interaction. The studies described here are the beginning of a research program designed to discover more about the role fragrance plays in social interaction. These exploratory studies are not experimental in nature; no attempt was made to confirm or disconfirm established theories or to test narrowly focused hypotheses. However, before formulating theories and designing studies to focus on specific hypotheses, more needs to be known about how and when fragrances are used, and what effects they have on social interaction. Our research addressed broad questions: How often do how many people use fragrances? Under what circumstances do people use fragrances? When do people think their fragrance has an impact on others? Experimental research is valuable, and will follow the exploratory studies described here.

This paper describes two investigations of fragrance use in social interaction. In both, participants maintained a diary of their social contacts using a variant of the Rochester Interaction Record (Wheeler & Nezlek, 1977). They maintained these diaries for two weeks, describing social contacts that lasted ten minutes or more. These diaries were updated once or twice each day, and we have every reason to believe that the diaries provided an accurate description of our participants' social lives. The diaries included descriptions of the duration of interactions, who was present, and participants' reactions to the events, including specific questions about the role that fragrances played in the interactions.

Method

Participants in the first study were 125 freshmen at the College of William & Mary. Participants in the second study were 90 freshmen and 85 juniors at the College of William & Mary. The juniors in the second study had all been participants in the first study. All volunteered in response to a request circulated in their classes, and all were paid $20 for their participation. A longitudinal design was chosen for two reasons. First, a second sample was needed to replicate the findings of the first study, and second, a longitudinal approach (comparing the same people over time) was needed to examine the stability of the findings over time.

In both studies, people described their social contacts along dimensions that past research has shown to be psychologically meaningful. These dimensions were: their enjoyment of the interaction, the influence they

FIGURE 1. *Interaction diary form used in Study 2.*

Scale used for all ratings for both studies:

1	2	3	4	5	6	7	8	9
not		slightly		somewhat		quite		very

Date_____   Time_____   A.M./P.M.   Length_____

People you were with:
                Initials: _____   _____   _____

Your   F/M   F/M   F/M
Reactions
_____
_____   Enjoyment   _____   _____   _____
_____   Influence   _____   _____   _____
_____   Closeness   _____   _____   _____
_____   Responsive   _____   _____   _____
_____   Confidence   _____   _____   _____
_____   Fragrance   _____   _____   _____
                                                 N__ P__ V__   N__ P__ V__   N__ P__ V__

Group: F   M   MX   No. _____   Activity _____

Note: All ratings were made on 1-9 scales.
       'N' refers to the importance of non-verbal information.
       'P' refers to the importance of para-verbal information.
       'V' refers to the importance of verbal information.

felt they had during the event, how intimate they felt the interaction was, how responsive others were to them, and how confident they felt. In addition, participants described their perceptions of the impact their fragrances had on the people with whom they were interacting. Impact was defined in two ways: Did you think that other people were aware of the fragrance you were wearing, and if they were aware of your fragrance, how much did they like it? Participants also described how they thought their co-interactants felt during the interactions, using the same dimensions they used to describe their own reactions. Detailed instructions for participants in diary studies are given in Nezlek and Wheeler (1984).

The important difference between the two studies was that in the second study participants also provided descriptions of the bases they used to make judgments about the impact of fragrance. Following established research on interpersonal communication and perception, participants described how much they relied on the following cues when they made judgments about how others were feeling or thinking: verbal cues (actual words), para-verbal cues (tone, volume, etc., the non-word parts of speech), and non-verbal cues (body language, facial cues, etc.). The diary form used in the second study is shown in Figure 1.

The interaction diary data were analyzed by programs written specifically to analyze data gathered for this project, and these procedures are summarized in Nezlek and Wheeler (1984). The pre-analysis programs calculated summary indices that described participants' social lives in three main categories. The overall index summarized all interactions. A second set of measures separately summarized interactions involving only co-interactants of the same sex, those involving only co-interactants of the opposite

sex, and those involving both men and women as co-interactants. A third set of measures summarized interactions with the best same- and opposite-sex friends of the diary keeper. Participants contributed equally to the final analysis regardless of their level of social activity.

The summary measures of social interaction were analyzed in three ways. One analysis compared mean differences among different types of interactions, e.g., differences in affective responses to same- versus opposite-sex interaction. A second analysis examined the relationships among different characteristics of interactions, e.g., whether individuals who experience more pleasure in interaction more socially active. A third strategy compared the interaction patterns of different groups of people, e.g. differences between males and females. All three strategies were used to analyze the data generated in the present study, although only selected results are presented here.

To facilitate the presentation of these results, details of the statistical procedures used to analyze the data are not presented. To test the differences of means between Studies 1 and 2, mixed model ANOVAs, with sex as a between factor and time as a within factor were conducted on the data describing those who participated in both Studies 1 and 2. To test the differences between the juniors and freshmen who provided the data in Study 2, two-way ANOVAs were conducted, with sex and academic year as between factors. Relationships among variables were examined with simple Pearson product-moment correlations. Correlations computed across time (relationships between Study 1 and 2 behaviors) involved only participants who were in both studies. In all cases, the commonly accepted significance level of .05 was used to distinguish random results from those with potential meaning.

The salience of fragrances in social interaction

Data describing how often participants felt that their fragrances were noticed by others are presented in Table 1. Two important trends emerged from the analyses of these data. First,

fragrances were more salient in opposite-sex than in same-sex social contacts. This difference was expected, and probably resulted from a variety of factors. For example, people may be

TABLE 1. *Percent of interactions in which fragrances were salient.*

| | Study 1 | | Study 2 | | | |
| | Freshman | | Juniors | | Freshman | |
Interactions	F	M	F	M	F	M
All	.26	.23	.17	.13	.09	.16
Same-sex	.19	.17	.14	.06	.07	.10
Opp.-Sex	.36	.33	.24	.20	.14	.25
Sample size	66	57	51	31	53	38

more likely to apply fragrance in anticipation of opposite-sex interactions than in anticipation of same-sex events, and opposite-sex contacts may be more physically intimate, providing a better opportunity for fragrances to be noticed.

Second, fragrances were rated as salient less frequently in Study 2 than in Study 1. This difference was not anticipated, and it cannot be explained by participants' reports of how they maintained their diaries. In both studies, participants were interviewed individually following completion of the diary, and they were asked specifically how they made judgments about how other people felt, including how they made inferences about the salience of their fragrance. The responses in these interviews did not suggest any awareness of the fact that the method used to measure inferences influenced reports of fragrance salience or favorability. Nonetheless, the differences in salience between the two studies may have been due to differences in the procedures used. In both studies, participants were instructed to indicate when they thought their fragrance was salient to others. However, in Study 2, they were also instructed to indicate the bases on which they made these judgments of salience. Perceived salience was generally lower in Study 2 than in Study 1. It is quite possible that participants' judgments about how often their fragrances were perceived by others were influenced by the additional attention they paid to thinking about how and why they believed others perceived these fragrances. That is, the extra attention paid to these social judgments seems to have made people more conservative when assessing the impact of their fragrances.

Moreover, given the responses to post-study interviews of the participants, and in focus groups conducted prior to the study about how often they applied fragrances, these lower estimates may be more realistic. Making judgments about the impact of one's fragrance is an uncommon task for most people. It may be that the novelty of making such judgments, free from providing a justification for them, lead to relative overestimates of salience in Study 1. In Study 2, although the judgments of fragrance impact were still novel, the request for a justification lead to more considered and more cautious judgments, in turn producing fewer situations in which impact was assumed. It is interesting to note however, that for people who were in both studies, the correlations between perceived fragrance salience across the two studies were high, suggesting that more detailed questions about perceptions of fragrance impact produce a uniform decrease in assumed salience.

The difference between the two studies in the reported salience of fragrance has implications for the study of fragrance in social interaction beyond studies using interaction diaries. The results suggest that judgments of the impact of one's fragrance are influenced by how these impacts are measured. Specifically, more detailed questions about fragrance impact lead to more conservative estimates of this impact. Further research is needed to determine the relative utility of requesting more or less detailed reports about fragrance impact.

TABLE 2. *Correlations between assumed favorability of fragrances and other characteristics of social interaction (all interactions combined).*

Characteristic	Study 1		Study 2	
	F	**M**	**F**	**M**
Enjoyment	.25	.42	.23	.40
Influence	.30	.30	.36	.47
Intimacy	.13	.40	.29	.43
Responsive	.09	.29	.35	.47
Confidence	.43	.47	.34	.51

Fragrance desirability and reactions to social interactions

Data describing participants' reactions to their social interactions, and how positive they thought others found their fragrance are presented in Table 2. The trend in these results is clear; the more positive that people thought others found their fragrance, the more positive they found their social interactions to be. Table 2 summarizes results describing all interactions taken together. When analyses were done separately for same- and opposite-sex interactions, similar patterns emerged, although the relationships were stronger for opposite-sex interactions than for same-sex interactions. The strongest and most relationships were found between assumed favorability and confidence. At this time it is not possible to speculate why this was so; it remains an interesting finding nonetheless.

Although the relationships between assumed favorability and other reactions to interactions were strong, they do not lend themselves to a straightforward interpretation. One possibility is that perceptions of fragrances and perceptions of other aspects of interaction are somehow causally related; another, less exciting possibility is that the relationships reflect a response bias: some people use the high ends of scales (no matter what the scale) and others use the low ends. Putting two such groups together would lead to a statistical relationship between two variables that would not reflect an underlying relationship between these variables. To control for this possibility, data from Study 2 were analyzed further.

Reactions to interactions in which fragrances were assumed to be salient were compared to reactions to interactions in which fragrances were not salient (Table 3). Analyses of these data suggest that interactions in which fragrances were assumed to be salient were affectively richer and more positive than interactions in which fragrances were not salient. This characterized both same- and opposite-sex interactions. Moreover, this was true not only for the reactions of people themselves to these events; it was true also for their perceptions of others reactions. A variety of plausible explanations exist for this phenomenon. One possibility is that people use fragrance during interactions that are inherently more enjoyable (dinner parties vs. raking leaves). Another possibility is that fragrance contributes in some way to the enjoyment, sense of confidence, etc. that occurs in social contacts. The differences between social interactions in which fragrances are salient, and those in which they are not, are the focus of future, more experimental and laboratory research that we hope to conduct.

TABLE 3. *Reactions to opposite-sex interactions: self-ratings and inferences of others' reactions (Study 2 only).*

	Self		Other	
Fragrance	**Y**	**N**	**Y**	**N**
Enjoyment	7.6	6.7	7.6	6.9
Influence	7.3	6.8	7.2	6.9
Intimacy	7.5	6.8	7.5	6.9
Responsiveness	7.4	7.0	7.5	6.9
Confidence	7.5	7.2	7.5	7.3

Note: Only interactions with one other person are included, and these represent approximately 70% of all interactions.

Fragrances and non-verbal cues in social interaction

In Study 2, people indicated the bases they used to make judgments about the reactions of others. In light of the difference in reactions between fragrance-salient and fragrance-nonsalient interactions, fragrance salience was examined for its importance to different channels of communication (cues). Relevant data are presented in Table 4. Analyses of these data indicated that non-verbal cues were more important in judging the reactions of others when fragrances were salient than when fragrances were not salient. Verbal and para-verbal cues were equally important across the two types of events. The two most plausible explanation for these results are that fragrances are more likely to be used in interactions in which non-verbal cues are inherently more salient (dinner parties vs. raking the leaves), and more interestingly, that fragrances may in some way trigger or create greater sensitivity to non-verbal cues. This second possibility will be the focus of more experimentally focused research in the future.

Summary

Our naturalistic diary studies of the role that fragrance plays in social interaction suggest that fragrances are positively related to the quality of social interaction, and that fragrances are somehow linked to the importance of nonverbal cues. Demonstrating these relationships was a first step, but an important one, in discovering how and why fragrances influence social life. Although we have no reason to believe that our student sample was composed of odd or abnormal people, there is no guarantee that the present results would hold in a different sample. This issue will need to be addressed. More importantly, the present results do not identify causal mechanisms. A variety of competing and plausible explanations exist for some of our more interesting results, and these must be examined in more controlled settings. The present studies have provided a good indication of what our future research should concern, and it is to this task that we will turn.

TABLE 4. *Importance of different sources used to infer reactions of others (Study 2 only).*

Fragrance	Non-verbal		Para-verbal		Verbal	
	Y	**N**	**Y**	**N**	**Y**	**N**
Same-sex	4.9	4.4	5.5	5.3	6.5	6.5
Opp.-Sex	6.5	4.8	5.6	5.5	6.3	6.7

Note: *Only interactions with one other person are included, and these represent approximately 70% of all interactions.*

References

Nezlek, J. B., & Wheeler, L. (1984). RIRAP: Rochester interaction record analysis package. Psychological Documents, 14, 6. (Ms. No. 2610)

Wheeler, L., & Nezlek, J. B. (1977). Sex differences in social participation. Journal of Personality and Social Psychology, 35, 742-54.

The Authors

John B. Nezlek

Dr. John Nezlek is a Professor of Psychology at the College of William and Mary. He is a social psychologist whose primary research interest is naturally-occurring social interaction. His specific interest in olfactory processes regards the influences fragrances and their perceived impacts have on social interaction.

Glenn Shean

Dr. Glenn Shean is a Professor of Psychology at the College of William and Mary in Williamsburg, VA. Dr. Shean is a clinical psychologist with wide research interests, including the relationship between the use of fragarnce products, mood variations and social relationships. He has worked as a management trainer and consultant to a number of large private and government organizations.

The Role of Olfactory Perception in Social Interaction

Mark Snyder

Mark Snyder
Mark Attridge
Department of Psychology
University of Minnesota

Abstract

We report the major findings of a four-year program of psychological research exploring the role of olfactory perception in social interaction. Our goal is to understand the strategic use of fragrances in creating and fashioning social images, and in maintaining and sustaining social roles. This research has included: (1) surveys of fragrance-related behaviors, attitudes, and motivations; (2) studies of how people respond to fragrance advertising; (3) tests of the brand loyalty of fragrance users; and (4) the development of a new research scale to measure individual differences in psychological involvement with fragrances.

The Role of Olfactory Perception in Social Interaction

Our program of research is concerned with the strategic use of fragrances in creating and fashioning social images, and in maintaining and sustaining social roles i.e., what psychologists call the *impression management* process. With respect to the impression management process, we addressed the following question: Can and do people systematically evoke particular patterns of social reactions and interaction outcomes by intentional use of fragrances?

With the support of the Olfactory Research Fund, we conducted several lines of inquiry to address this question: (1) surveys of fragrance behaviors, attitudes, and motivations; (2) studies of how people respond to fragrance advertising; (3) experiments and surveys on the brand loyalty of fragrance users; and (4) the development of a new research scale to measure individual differences in psychological involvement with fragrances.

The theoretical context for this work is provided by a large body of theory and research in social psychology. This work suggests that people (and some people more than others) employ a wide variety of strategies and tactics, some subtle and others not so subtle, to control the images they convey to others with whom they interact (Snyder, 1987, 1989). Although much is known about the processes of *impression management*, the present studies are, to our knowledge, the first to systematically address the role of fragrance use in such processes.

I. Surveys of Fragrance Behavior, Attitudes, and Motivations

An important first step in our research was a major survey of behaviors, attitudes, and motivations relevant to the role of fragrance in social interaction (Snyder, 1990). A sample of 286 young adult men and women (university students) reported on their patterns of fragrance use, including their motivations for using fragrances, the situations in which they use fragrance, and the linkages between their fragrance use and various attitudinal and personality factors.

As a behavior, fragrance use is widespread among young adults. In our sample, 94% of women and 77% of men wear fragrance; in addition, only 3% of women and 8% of men claim to use no fragrance at all. As expected, women use fragrances more frequently, and own more fragrance products than men. Further, we found that although women typically use fragrances in a wide variety of social situations (at work, at school, with friends, on a date, etc.), men most often wear fragrance in situations involving women (e.g., on a romantic date) and far less often in other kinds of situations.

Regarding attitudes about fragrance, we found that our respondents could be typed meaningfully into two categories of fragrance users. In one category are people who use fragrance for fashioning and controlling their social images. They strategically select fragrances to help them display a specific social image for different social occasions. These people employ a wide range of fragrances, each chosen for its appropriateness to one of the specific roles of their lives: one fragrance for work situations, another for romantic situations, still another for family situations, etc. Personality measures reveal that these people are what are known as *high self-monitors* in many domains of their lives; such people are highly invested in monitoring or controlling the social images they project (Snyder, 1987).

In the second category are people who use fragrances not to create images and play roles, but to express their personalities. These consumers choose their scents on the basis of their personal reactions to the fragrances, not on the basis of image. In addition, they search for one fragrance that captures the essence of their identity, and they employ it across a wide range of situations, roles, and relationships. Personality inventories indicate that these people are *low self-monitors.* In many domains of their lives they seek to be true to their own sense of self, and they look for ways to display their true personalities in dealings with other people (Snyder, 1987).

Motivations for using fragrance also differed among individuals. For example, we constructed a scale from a subset of the items that measure the extent to which respondents' reasons for using fragrances reflects considerations of enhancing their social images. On this scale, high self-monitors are substantially more likely to use fragrances for impression management purposes than are low self-monitors.

To determine the reliability of these findings, we conducted a second survey of over 250 college students. Again, we found very high levels of fragrance use among women and men, and we obtained the same pattern of results for the differences between high and low self-monitors in their attitudes and motivations associated with fragrance use. As a general result, our surveys support the theoretically based expectation that people differ systematically in their patterns of fragrance use, and that they can be typed meaningfully into two categories.

FIGURE 1. *Image-oriented ads*

SUCCESS HAS ALWAYS BEEN YOUR STYLE

Male

TIMELESS ROMANCE

Female

II. Advertising and Marketing of Fragrance Products

Having found differences among people in their attitudes and motivations underlying fragrance use lead us to ask the question: How should one market fragrance products to each type of user? In our research, we have found that high self-monitors are particularly sensitive to image considerations in advertising, whereas low self-monitors are particularly sensitive to information about the fragrance product itself. Consider one study in which we created magazine-type ads for perfumes and colognes (Snyder & Attridge, 1993). In our ads, both the picture and the written message conveyed information <u>either</u> about the image associated with the fragrance and its users, <u>or</u> about the qualities of the fragrance product itself.

Thus, an ad for a women's perfume featuring a picture of a couple in a romantic setting with the slogan "Timeless Romance" would constitute an image-oriented pictorial message and an image-oriented written message. In contrast, an ad featuring a picture of the fragrance product itself, and the slogan "A Soft Floral Scent With a Hint of Musk," constitutes a product-oriented picture and a product-oriented written message. Similarly, an ad for a men's cologne featuring a young man who is the picture of success and upward mobility, and the slogan "Success Has Always Been Your Style," would constitute an image-oriented pictorial message and an image-oriented written message. An ad featuring a picture of the fragrance product itself and the slogan "A Fresh Spicy Blend of Citrus and Jasmine" consists of a product-oriented picture and a product-oriented written message. Although we used professional-looking ads in the study, Figure 1 offers simple illustrations depicting the image-oriented ads, and Figure 2 shows illustrations of the product-oriented ads.

FIGURE 2. *Product-oriented ads*

A FRESH SPICY BLEND
OF CITRUS AND JASMINE

Male

A SOFT FLORAL SCENT
WITH A HINT OF MUSK

Female

When college students evaluated these ads, high self-monitors assigned more favorable evaluations to the ads that convey information about the images to be gained by using the fragrance; low self-monitors assigned their most favorable evaluations to those ads that convey information about the fragrance product itself. Consider the marketing implications of these findings. They suggest separate advertising campaigns may be required to reach these two types of consumers: an image-oriented campaign to appeal to consumers high in self-monitoring, and a product-oriented campaign to appeal to consumers low in self-monitoring.

III. Brand Loyalty of Fragrance Users

The findings from the advertising study may also say something about brand loyalty. They may help us understand why some consumers (low self-monitors, our research would suggest) become habitual and loyal users of one brand (presumably one they believe captures the essence of their personality and identity), and why other consumers (high self-monitors) are less loyal to any one brand, and instead use a variety of brands (each, presumably, relevant to some role domain of their lives). To examine these implications concerning brand loyalty, we conducted both an experiment and a survey study.

In the *experimental study*, we sought to examine how different personality types demonstrated their brand loyalty (or lack thereof) to various fragrance products in a laboratory setting, where the type of scent and the images of the fragrances could be controlled (Haugen & Snyder, 1989). In designing this study, we assumed that people who want to wear image-appropriate fragrances in different situations would necessarily own and wear different fragrances, thus making them less loyal to one particular brand of fragrance. In contrast, people who are not interested in image-situation matching, but are interested in choosing perfumes based on scent and personal preferences, would own fewer perfumes and thus be more brand loyal. We predicted that when given a choice of different perfumes to wear, and a variety of situations in which to wear them, high self-monitors would be more likely than low self-monitors to match the image of a perfume with the specific situation in which they would wear it.

In one-on-one interview sessions, 51 college women followed a procedure in which they smelled four perfumes (each one a distinct type of scent: herbal, floral, citrus, and musk), saw the magazine advertisement corresponding to each perfume (each ad conveying a distinct image: business career, sporty, sophisticated/ cultured, and romantic), and were asked which of the perfumes she would be most likely and least likely to wear in four different situations. In each situation there was a perfume that could be considered most appropriate to wear in terms of its advertised image. For example, the most appropriate perfume to wear to an athletic event would be a perfume advertised with a sporty image.

The results showed that the high self-monitors were in fact more likely to choose the perfume that matched the social situation than were low self-monitors. These results suggest that, when compared to low self-monitoring women, high self-monitoring women were more attuned to the images with which products were advertised, and that they used this information when selecting brands to wear for different hypothetical situations. High self-monitors were interested in being the right person for each situation, and fragrance products were another tool to use. The desire of high self-monitors to match perfume images with situations relates to brand loyalty: by choosing to own and wear multiple perfumes to match different images and situations, they are necessarily less brand loyal. However, the product choices of low self-monitors appeared to be driven more by personal preferences than by product images. The desire of low self-monitors to choose perfumes based on personal preferences may lead them to use

fewer products, and to remain loyal to the ones they like.

In the *survey study*, we set out to identify consumers who choose to wear one fragrance brand for all situations, and those who wear different fragrances for different situations in everyday life (Haugen, Snyder & Attridge, 1990). To accomplish this goal, we created and administered a self-report questionnaire in which respondents could identify themselves as being brand loyal to fragrance products, or as lacking in brand loyalty. We predicted that low self-monitors would be more loyal to the fragrances they use than would high self-monitors, and would reveal this brand loyalty by choosing to use predominantly one fragrance. In addition, we predicted that when asked about their motivations for being brand loyal or non-brand loyal, brand loyal users would be more likely to generate reasons such as product quality, choosing their fragrances based on personal preferences, and choosing the one that pleases them, independent of the image that it may convey. In contrast, we predicted that multiple brand users would be more likely to generate reasons related to the images of the products, choosing their fragrances based on the multiple images they can project by using these products.

Analyses of the results of this survey, from a sample of nearly 300 young adult men and women, supported our predictions. For example, men who typically used only one fragrance tended to have low self-monitor personality type, whereas men who typically used multiple fragrances tended to be high self-monitors. In addition, those who said they were loyal to one fragrance brand gave reasons such as: "I like that brand's scent," "I only want one scent to be associated with me," or "One product is more convenient." Multiple product users, or non-brand loyal users, gave reasons such as: "I like to have different fragrances for different situations, "I like to wear different fragrances depending on my mood," or "I like to wear different fragrances depending on with whom I am going to spend time." Thus, the reasons given for being brand loyal were very different from those given by multiple product users.

Brand loyal users gave reasons related to the quality or distinctiveness of the products they wore, whereas multiple brand users gave reasons based on more transient and shifting aspects such as situations and moods. In addition, non-brand loyal fragrance consumers scored higher on a six-item measure of using fragrance for social image-oriented motives than did brand loyal consumers. Thus, as predicted, the non-brand loyal respondents wore fragrance products for more social and/or image reasons than the brand loyal users.

IV. The Construct of Fragrance Involvement

The most recent phase of our research program involves the articulation of a new construct of considerable potential significance in understanding the psychology of fragrance. As we have contemplated the implications and significance of our line of research on fragrance and social behavior, we have become increasingly convinced that, underlying our findings, there exist differences between people even more fundamental, pervasive, and consequential than those we have identified thus far. These differences concern the basic processes of how individuals think about and respond to fragrances and fragrance-based information, and the extent to which fragrance is meaningfully involved in diverse domains of people's functioning as individuals and as social beings.

We conceive of these differences as a new psychological construct that we refer to as *fragrance involvement.* We believe that this construct, and measures of it that we have developed and validated, have the potential to integrate, synthesize, and systematize diverse aspects of the role of fragrance in people's thoughts, feelings, and actions. The fragrance involvement construct has four components: *interest* in fragrance related issues, *attention* to information about fragrance related issues, *emotional responsiveness* to fragrance, and *behaviors* related to fragrance. We developed a 16 item self-report scale that assesses the four conceptual components of fragrance involvement. Sample items from the scale include: "How interested are you in fragrances?," "How often do you notice fragrances that other people are wearing?," "How extreme are your emotional reactions to fragrances?," and "If someone is wearing a fragrance that you like, how likely is it that you would inquire about the name of it?." In developing the scale, we started with a survey of 89 young adult females, and conducted a follow-up survey with 50 female and 50 male college students.

Results of the two surveys showed that high fragrance involvement was associated in expected ways with a wide array of important consumer behaviors, attitudes, and motivations. Specifically, high fragrance involvement was associated with the following: frequent use of fragrances, owning and using a large number of different fragrance products, carrying a fragrance on their person, keeping fragrances at work or school, a greater willingness to "try-out" new fragrances, devoting time to shop for fragrances, buying fragrance products at several stores, spending more money on fragrances, remembering a greater number of specific brands of fragrances, remembering a greater number of specific magazine and TV advertisements for fragrances, remembering more information or news about fragrances, remembering a greater number of specific personal experiences involving fragrance, use of fragrance for reasons of social and image-oriented motives, greater use of cosmetics (for women), greater awareness of clothing fashion trends, and being a high self-monitor personality type. It is important to note that females were higher in fragrance involvement than were males. In sum, these findings offer empirical evidence that the construct of fragrance involvement can be reliably assessed, and that is related to a host of fragrance-related attitudes and behaviors.

Conclusions

Based on this program of research, it seems that fragrance users can be meaningfully typed into two categories. In one category are people who in addition to liking its scent wear fragrance because it helps them fashion and control their social image, strategically choosing fragrances to help them display specific images for a specific social occasion. These people use multiple fragrances, usually chosen for the appropriateness of each to a specific role they play in their lives. They like the image that the fragrance helps them project in that role. These people tend to be high self-monitors, people who are highly invested in monitoring and controlling the images they project. Fragrance products, it would seem, are another prop for them to use in their impression management activities.

In the other category are people who use fragrances, not necessarily to create images and play roles, but because they like the scent quality of the particular product. In addition, they search for one fragrance that captures the essence of their identity, a scent they like and can cross a wide range of situations, roles, and relationships. These people tend to be low self-monitors, people who, in many domains of their lives, seek to be true to their sense of self and look for ways to display their true personality in dealing with others. For these people, fragrances may be instrumental in expressing their personalities in a consistent and coherent fashion across circumstances.

Lastly, the introduction of our new construct promises to classify people as being high or low in fragrance involvement. Level of fragrance involvement should be systematically related to a number of meaningful aspects of fragrance use and social behavior, which would be of particular interest to both marketers of fragrance products and researchers. For example, high fragrance involvement should be associated with increased use of fragrance products and with a more complex and elaborate understanding of fragrance. As a function of this increased sophistication, higher involvement should be associated with a broader range of motivations and reasons for fragrance use, and even greater likelihood of fragrance-based reactions to others. Our future research plans involve further validation of the fragrance involvement construct and self-report measure.

References

Haugen, J. & Snyder, M. (May 1989). "Who is the brand loyal consumer?" Research paper presented at the Midwestern Psychological Association annual meetings, Chicago, IL.

Haugen, J., Snyder, M. & Attridge, M. (May 1990). "Self-monitoring and the brand loyal consumer." Research paper presented at the Midwestern Psychological Association annual meetings, Chicago, IL.

Snyder, M. (1987). Public appearances/private realities: The psychology of self-monitoring. New York: Freeman.

Snyder, M. (1989). Selling images and selling products: Motivational foundations of consumer attitudes and behavior. In T.K. Srull (Ed.), Advances in consumer research, 16, 306-11. Provo, UT: Association for Consumer Research.

Snyder, M. (1990). Fragrance and social behavior. Perfumer & Flavorist, 15, 37-38.

Snyder, M., & Attridge, M. (1993). How personality affects reactions to advertising for fragrance products. Fragrance Forum, 9 (4), 6.

The Author
Mark Snyder

Dr. Mark Snyder is Professor of Psychology and Chair of the Department of Psychology at the University of Minnesota, where he has been a member of the faculty since 1972. His research interests include theoretical and empirical issues associated with the motivational foundations of individual and social behavior, and the applications of basic theory and research in personality and social psychology to addressing practical problems confronting society. He has served as President of the Society for Personality and Social Psychology. He is also the author of the book, Public Appearances/Private Realities: The Psychology of Self-Monitoring.

Mood

physiology • olfactory co

nd social behavior • mood • cultural and historical perspectives • applications • fundamentals of odor per

and psy

V

nentals of

perception • fragrance and psycho

There is an emotional component in our response to scent, and this suggests that fragrance might affect psychological processes in much the same way that mood does. This proposition has been tested in studies by <u>Howard Ehrlichman</u>, using pleasant and unpleasant odors. He found mood-like effects of odor on complex behavior (for example, the recollection of memories), and on simple, automatic behavior (the startle reflex). These studies provide solid experimental evidence that scent affects mood.

A natural place to examine the potential effects of fragrance on mood is in people experiencing the biological and psychological stress of middle age. Menopause in women, and aged-related physiological changes in men, can produce a variety of physical and mental symptoms. <u>Susan Schiffman and Elizabeth Sattely-Miller</u> examined the efficacy of pleasant scents in improving mood in this population, and found encouraging results.

Influence of Odors on Mood-Related Behavior

Howard Ehrlichman
Queens College and the Graduate School
of the City University of New York

Abstract

Our research program has focused on the relationship between odor and mood. A great deal of research has demonstrated that people's moods influence how they think, what they remember, and how they perceive others. In a series of studies, we attempted to see if odors could mimic the effects of mood states; that is, we asked whether smelling a pleasant or unpleasant odor is psychologically similar to being in a good or bad mood. This report describes our research involving memories, optimism, thoughts about helping, evaluation of ambiguous stimuli, creativity and psychophysiological reactivity. Although not all of our results have been supportive, the general picture that emerges from this research is that odors can produce psychological effects similar to mood states.

Influence of Odors on Mood-Related Behavior

Many of us believe that lovely fragrances and foul stenches can influence our feelings. The emotional impact of scents is central to perfume use, aromatherapy, and product and environmental fragrancing. Nevertheless, until quite recently there has been little scientific attention paid to this fascinating subject by research psychologists (Ehrlichman & Bastone, 1992a). As a result, many ideas about the relationship between olfaction and emotion have been neither precisely formulated nor subject to empirical investigation. Furthermore, most of what people believe about the emotional impact of odors comes not from scientific evidence, but from personal experience, anecdotes, or commonly held opinion. But how far can we trust such informal evidence? Does smelling a favorite fragrance really improve our mood? Does living near a sewage treatment plant that produces noxious odors really make our mood worse?

One way we try to answer this question is to ask people how they feel when they smell different fragrances. Unfortunately, such self-reports are often influenced by factors such as how questions are asked, what people think the questioner wants to hear, and so on. As a result, we wished to go beyond what people say and look at what they do. Psychologists who study emotions have discovered that moods can influence people's behavior and thoughts (Isen, 1984, 1990). For example, studies show that if people are asked to recall memories from their past, people in a good mood remember more happy experiences as compared to people in a bad mood (Teasdale & Fogarty, 1979). As another example, people in a good mood appear to think more creatively than people in a bad mood (Isen, Daubman & Nowicki, 1987).

Overview of research design

Our research has been designed to find out if the sense of smell can change people's behavior and thought processes just the way good moods and bad moods do. In other words, we have attempted to go beyond people's reports of how fragrances make them feel to see if there is a real shift in their psychological functioning when they smell pleasant or unpleasant odors.

We have carried out a series of studies taking this approach. These studies take the same general form. Subjects come to the lab and are presented with pleasant or unpleasant odors while they engage in a mood-sensitive task. We have used a variety of different odorants in this research. Our pleasant odors have included both food (e.g., almond and coconut) and floral (e.g., muguet and water lily) scents. Our unpleasant odors have included chemicals (e.g., pyridine and butyric acid) and natural substances (e.g., limburger cheese and cigar butt). In some studies, we also included a no-odor control condition. In all studies, subjects rated the odors for pleasantness and unpleasantness, and we only included those subjects who responded to the odors as intended.

Effects on personal memories

What is currently in our minds influences what we are likely to remember. If we are thinking about insects, and someone utters the word "fly," we are more likely to think of a housefly than an airplane. Similarly, if we are in a good mood, memories that are associated with positive feelings are more likely to come to mind than memories associated with negative feelings (Bower, 1981). In our first study (Ehrlichman & Halpern, 1988), we wanted to see if presenting pleasant and unpleasant odors to people would also influence what memories would come to mind. Would the pleasantness or unpleasantness of an odor experience be enough to bias people's memories toward happy or unhappy memories? In order to test this idea, we asked people to think of the first memory from their past that came to mind in response to 20 neutral words (such as table) that we presented over audio tape. This was done while subjects smelled either a pleasant odor, an unpleasant odor, or no odor. After recalling 20 memories, subjects rated each memory with regard to how happy or unhappy it was at the time it happened. We found that merely smelling pleasant or unpleasant odors influenced the memories people recalled. People who smelled the pleasant odor of almond reported relatively more happy memories, and fewer unhappy memories, than those who smelled the unpleasant odor of pyridine. In other words, odors seemed to influence memories just as moods do: a pleasant odor is more likely to trigger happy memories than is an unpleasant odor, and vice versa.

Risk estimates and optimism

Some investigators have reported that people in a good mood think the world is a safer place (Johnson & Tversky, 1983) and expect more positive future events (Forgas & Moylan, 1988) than do people in a bad mood. Accordingly, we had subjects guess the number of people in the United States who would die during the next year from various causes (such as tornadoes) and rate the probability of future good and bad events (e.g., you are very successful in your work; your car is stolen) while smelling either pleasant or unpleasant odors. We found no influence of odors on probability estimates (Ehrlichman & Bastone, 1992b).

Evaluations of ambiguous stimuli

Moods seem to influence how we see the world. In good moods, the world seems brighter and we like people more; in bad moods, all seems dark and we often get annoyed and irritated with people. When people are asked to say whether they feel positively or negatively about neutral words or individuals they see in photographs, people in good moods are more positive in their evaluations than are people in bad moods (Isen & Shalker, 1982; Kuykendall, Keating & Wagaman, 1988). Rotton (1983) had subjects evaluate paintings, people in photographs, and descriptions of people while exposed to an unpleasant odor (ethyl mercaptan) or with no odor present. There was some evidence for more negative evaluations in the unpleasant odor condition. In contrast, Cann and Ross (1989) found that men's ratings of the attractiveness of women shown in slides was not influenced by pleasant or unpleasant odors.

In order to explore the effects of odors on evaluations, we presented subjects with slides showing individual words or pictures of people that were previously rated as neither clearly positive nor negative. Subjects evaluated these stimuli once during exposure to a pleasant odor and once during exposure to an unpleasant odor. Subjects were asked to rate how positive the words were based on the first thing they brought to mind. For example, if the word was "cat" and the first thing brought to mind was one's favorite pet, the rating would be positive. If the first thing brought to mind was getting scratched, it would be negative. For the pictures, subjects were asked to rate physical attractiveness and whether the person in the photograph appeared to have a pleasant or unpleasant personality.

Because previous research on the effects of odors on evaluations had been inconsistent, we thought it might be worthwhile to see if

characteristics of subjects might be important. Accordingly, subjects were given a test of a cognitive-personality characteristic called "field dependence" (Witkin & Goodenough, 1981). Because field-dependent people are more likely to be influenced by external stimuli than are field-independent people, we thought that the effects of odors on evaluations of words and photographs might be greater for field-dependent people.

The results supported this idea. Field-dependent people rated the words and the personalities of the people in the photographs more positively when they smelled pleasant as compared to unpleasant odors. In contrast, there was no effect of odors on field independent subjects' ratings of words or personalities. No effects of odors were found for either group of subjects on ratings of phsycial attractiveness.

Creativity

Isen, Daubman, and Nowicki (1987) reported that good moods improved people's performance in tasks involving creativity. In order to see if odors could also influence creativity, we conducted two studies in which subjects performed creative tasks while they smelled a pleasant odor (almond or muguet) or an unpleasant odor (butyric acid or thiophene). One task (the Remote Associates Test), involved coming up with a word that ties other, seemingly unrelated words, together (for example, CLUB GOWN MARE). A second task (the Alternate Uses Test) asked subjects to come up with novel uses for familiar objects (for example, how many things can you do with a brick). Both of these tasks have been used as measures of creativity in previous research. In the first study, which used just the Remote Associate Test, subjects who smelled a pleasant odor did better than those who smelled an unpleasant odor. This result was quite similar to that reported by Isen et al. (1987), suggesting that good and bad odors can produce changes in people's thought processes that are similar to the effects of good or bad moods.

The second study employed both the Remote Associates Test and the Alternate Uses

Test (Bastone, 1992). To our surprise, this time we found no effect of odors on the Remote Associates Test. However, there was a significant effect of odors on the Alternate Uses Test: subjects in the unpleasant odor condition came up with significantly fewer uses (that is, performed less creatively) than subjects in the pleasant or no-odor conditions.

The reason for the difference in the findings with the Remote Associates Test in not obvious. There were a number of changes in the way the test was administered between the first and second studies that might account for the discrepancy. Without additional research it is difficult to know whether this interpretation is correct. In any case, we view the overall results as generally supporting the idea that odors can influence performance on creativity tests.

Physiological reactivity: startle modulation

The studies we have reviewed to this point all involved very complex behaviors. Although they were chosen because previous research had found them to be influenced by mood states, it is also true that the effect of mood on such behaviors is not usually very strong and sometimes difficult to replicate. There can be little doubt that optimism, thoughts about helping, and creativity are influenced by many factors other than mood. Given the complexity of these behaviors, it is not surprising that the effects of pleasant and unpleasant odors tend to be small and difficult to obtain. Indeed, the fact that odors can sometimes produce significant changes in peoples' memories, creativity and evaluations is, in our view, rather remarkable. At the same, time, the very complexity of these behaviors may limit their usefulness for answering the question that we set out to address: do pleasant and unpleasant odors produce psychological changes similar to those that accompany positive and negative mood states?

An alternative approach would be to more directly measure changes in psychological state using psychophysiological techniques. Our last study in this series took this approach, using a phenomenon known as "startle reflex modulation" (Ehrlichman, Brown, Zhu & Warrenburg, 1993). When people hear a sudden loud noise, they respond with a "startle reflex." A great deal of research has shown that the size of this startle reflex depends on the mood a person is in (Lang, Bradley & Cuthbert, 1990). When people are thinking about pleasant events or looking at pleasant pictures, the startle is reduced. When they are thinking about unpleasant events or looking at unpleasant pictures, the startle increases. We wanted to see if smelling pleasant and unpleasant odors would have similar effects. Subjects were asked to sniff a series of bottles containing either pleasant odorants, unpleasant odorants, or no odor. While they sniffed, loud, sudden noises (like pistol shots) were presented over earphones. We measured the startle reflex by monitoring muscle activity (using electromyography) produced by the eyeblink that is the primary motoric component of the startle reflex. The results were as follows: when people smelled unpleasant odors (like limburger cheese), sudden loud noises produced larger startle reflexes as compared to the reflexes in a no-odor control condition. However, there was no effect of pleasant odors on the size of the startle reflex, despite the fact that the pleasant odors (like coconut) were rated just as pleasant as the unpleasant odors were rated unpleasant.

These results offer further support that odors can produce psychobiological changes in people similar to those linked to moods. It is noteworthy that in this study the response was a simple, non-voluntary reflex. Taken together with the results of our previous research, we feel confident in concluding that odors can have real effects on people's emotional states. At the same time, it is obvious that the effects of odors are not simple. One cannot assume that any positive or negative odor will necessarily influence mood without taking other factors into account. This seems to be particularly true for pleasant odors. Although in some of our studies pleasant odors did seem to influence behavior, more often it was the unpleasant odors that seemed to have the stronger effects. This asymmetry between pleasant and unpleasant odors was most evident in the startle reflex study, but it was also observed when we simply monitored self-reports of mood states as people experienced different odors (Ehrlichman & Bastone, 1992b). Why might the effects of unpleasant odors on mood and mood-related behavior be stronger than the effects of pleasant odors? One possibility is that unpleasant odors have a more direct and immediate impact on our feelings than do pleasant odors. It is important to emphasize that in all of these studies, people are experiencing the odors in a laboratory setting, and the odors are coming from bottles or plastic tubes. Perhaps the emotional power of pleasant odors is strongest when they are experienced in a context that promotes a generally positive feeling state. A fragrance sniffed from a bottle may simply not have the same impact as one emanating from an attractive person or a lovely garden. In contrast, an unpleasant odor may not require a context in order for it to have a strong impact on a person's feelings and mood state.

In conclusion, our research strongly suggests that the effects of odors on people's emotional states are real and can be demonstrated in rigorous, scientific studies. We suggest that these effects will be stronger when odors are experienced in appropriate contexts, and this will be especially true for pleasant odors.

References

Bastone, L.M. (1992). Odor experience as an affective state: Effects of odor pleasantness on creativity. Unpublished doctoral dissertation, City University of New York.

Bower, G.H. (1981). Mood and memory. American Psychologist, 36, 129-48.

Cann, A. & Ross, D.A. (1989). Olfactory stimuli as context cues in human memory. American Journal of Psychology, 102, 91-102.

Clark, M.S. & Waddell, B.A. (1983). Effects of moods on thoughts about helping, attraction and information acquisition. Social Psychology Quarterly, 46, 31-5.

Ehrlichman, H. & Bastone, L. (1992a). Olfaction and emotion. In M.J. Serby & K.L. Chobar (Eds.), Science of olfaction, 410-38. New York: Springer-Verlag.

Ehrlichman, H. & Bastone, L. (1992b). The use of odour in the study of emotion. In S. Van Toller & G.H. Dodd (Eds.), Fragrance: The psychology and biology of perfume, 143-60. London: Elsevier.

Ehrlichman, H., Brown, S., Zhu, J. & Warrenburg, S. (1993). Startle reflex modulation during exposure to pleasant and unpleasant odors. Manuscript submitted for publication.

Ehrlichman , H., & Halpern, J.N. (1988). Affect and memory: Effects of odor pleasantness on cognition. Journal of Personality and Social Psychology, 55, 769-79.

Forgas, J.P., & Moylan, S. (1988). After the movies: Transient mood on social judgements. Personality and Social Psychology Bulletin, 13, 467-77.

Isen, A.M. (1984). Toward understanding the role of affect in cognition. In R.S. Wyer, Jr. & T.K. Srull (eds.), Handbook of social cognition, 179-236. Hillsdale, NJ : Lawrence Erlbaum.

Isen, A.M., Daubman, K.A. & Nowicki, G.P. (1987). Positive affect facilitates creative problem solving. Journal of Personality and Social Psychology, 52, 1122-31.

Isen, A.M. & Shalker, T.E. (1982). The effect of feeling state on evaluation of positive, neutral, and negative stimuli: When you "accentuate the positive," do you "eliminate the negative?" Social Psychology Quarterly, 45, 58-63.

Isen, A.M. (1990). The influence of positive and negative affect on cognitive organization: Some implications for development. In N.L. Stein, B. Leventhal & T. Trabasso (Eds.), Psychological and biological approaches to emotion, 75-94. Hillsdale, NJ: Erlbaum.

Johnson, E.J., & Tversky, A. (1983). Affect, generalization, and the perception of risk. Journal of Personality and Social Psychology, 45, 20-31.

Kuykendall, D., Keating, J.P., & Wagaman, J. (1988). Assessing affective states: A new methodology for some old problems. Cognitive Therapy and Research, 12, 279-94.

Lang, P.J., Bradley, M.M., & Cuthbert, B.N. (1990). Emotion, attention, and the startle reflex. Psychological Review, 97, 377-98.

Rotton, J. (1983). Affective and cognitive consequences of malodorous pollution. Basic and Applied Social Psychology, 4, 171-91.

Teasdale, J.D. & Fogarty, S.J. (1979). Differential effects of induced mood on retrieval of pleasant and unpleasant events from episodic memory. Journal of Abnormal Psychology, 88, 248-57.

Witkin, H.A., & Goodenough, D.P. (1981). Cognitive styles: Essence and origins. New York: International Universities Press.

The Author

Howard Ehrlichman

Dr. Howard Ehrlichman is a Professor of Psychology at Queens College and the Graduate School of the City University of New York, serving in the doctoral programs in Social-Personality and Neuropsychology. A psychologist who studies personality, individual differences, brain-behavior relations and emotion, Dr. Ehrlichman is also the author of two book chapters on the topic of olfaction and emotion.

Pleasant Odors Improve Mood of Women and Men at Mid-life

Susan S. Schiffman
Elizabeth A. Sattely-Miller
Departments of Psychology and Psychiatry
Duke University

Physiological and psychological changes that can interfere with positive mood occur in both women and men at mid-life. In women, the climacteric generally occurs between 41 and 59 years of age and marks the transition from the reproductive to nonreproductive state (Barbo, 1987). It is characterized by endocrine, somatic, and psychological symptoms including hot flashes, night sweats, insomnia, nervousness/irritability, joint pains, headache, fatigue, depression, dizziness, difficulty in concentrating, nervousness, and a vague feeling of unwellness (Barbo, 1987; Flint, 1975; Padwick, et al., 1985; Polit & La Rocco, 1980). A male climacteric has also been described in the medical literature and is a life transition characterized by biological and psychological aspects (Featherstone & Hepworth, 1985; Vartapetov & Demchenko, 1965; Parlee, 1978).

Both the biological and psychological aspects of climacteric in women and men have been attributed in part to hormonal changes. In women, cessation of estrogen production (Hunter, 1992; Vliet & Davis, 1991) contributes to physiological and psychological complaints; estrogen treatment has been found to be helpful in treating these mood swings (Padwick, et al., 1985; Wiklund, et al., 1992; de Lignieres & Vincens, 1982). Conversely, some somatic symptoms of menopause including palpitations, dizziness, headaches, fatigue, insomnia, irritability, and weight gain are attributed to psychological dysfunction from mid-life problems (Holte & Mikkelsen, 1991; Schindler, 1987; Schmidt & Rubinow, 1991; Kalmar, et al., 1992). Physiological changes have also been documented in some males during mid-life, including diminished testosterone (Szarvas, 1992; Janczewski, 1967; Greenblatt, et al., 1979; Soules & Bremner, 1982), increased levels of the carrier protein sex hormone-binding globulin (SHBG) which results in less androgen supply (Polit & La Rocco, 1980), enhanced serum luteinizing hormone

(LH) levels (Szarvas, 1992; Soules & Bremner, 1982), decreased spermiogenetic activity of the testes (Janczewski, et al., 1967; Nankin, 1985), and degenerative changes of the penis (Nankin, 1985). Androgen replacement therapy can improve libido (Greenblatt, et al., 1979; Jacobelli, 1985) and may also have psychological benefits including reductions in fatigue, depression and headaches (Greenblatt, et al., 1979) . For men the biological changes at mid-life are gradual, while for women the biological changes are relatively abrupt over a 1-2 year span (Nankin, 1985).

In addition to the biological changes that occur during climacteric, the psychological issues during mid-life can have equal if not greater impact. In women, factors such as stages of family development, personal identity issues, and societal expectations are very important (Schindler, 1987). Males often experience inner turmoil, altered aspirations, career frustration, confrontation with death, family/role changes, and concern with a decline in sexual potency (Holte & Mikkelsen, 1991; Schindler, 1987; Schmidt & Rubinow, 1991; Kalmar, et al., 1992).

Recent experiments have reported that pleasant odors can elevate mood of college-age women. Ehrlichman and Bastone (1992) found pleasant and unpleasant odors differentially affected mood. Pleasant odors improved mood while unpleasant odor induced feelings of disgust. In another study, Ehrlichman and Halpern (1988) reported that pleasant odors produced happier memories than unpleasant odors in college-age women.

The purpose of the present investigation was to determine experimentally if the odor of fragrances can improve the mood of women and men at mid-life since previous studies show that fragrance improves mood in college-age women. Two separate studies were conducted which used the Profile of Mood States (POMS) questionnaire (McNair & Lorr, 1964; McNair, et al., 1992) to assess the effect of fragrance on mood states. The POMS (McNair & Lorr, 1964; McNair, et al., 1992) was used to measure the impact of the fragrances because it has been extensively tested and validated, and because it has been shown to be sensitive to transient mood shifts (Galil, et al., 1990; Cockerill, et al., 1991; Cole, et al., 1978; Der & Lewington, 1990; File, et al., 1982; Horswill, et al., 1990; Kraemer, et al., 1990; Lieberman, et al., 1982/83; Rausch, et al., 1990; Williams, et al., 1991; Kantor, et al., 1978).

Methods

In the first study, the effect of perfumes upon mood was tested in fifty-six women ranging in age from 45 to 60 years. All female subjects were either Duke University employees or worked within the Durham, North Carolina community. The fifty-six female subjects were divided into four groups of fourteen subjects each on the basis of hormonal status. Group 1 consisted of women who were still menstruating; group 2 consisted of women who were no longer menstruating and taking estrogen; group 3 consisted of women who were no longer menstruating and taking estrogen and progesterone; and, group 4 consisted of women who were no longer menstruating and taking no hormone replacement. The date that each woman began study 1 was coordinated with hormonal status. Women in group 1 entered the study on the first day of their menstrual cycles; women in group 2 and group 3 entered the study on the first day of their estrogen cycles; and, women in group 4 could enter the study at any time.

In the second study, sixty men were recruited from the Durham, North Carolina area to participate. These men, who were between the ages of 40 and 55, were divided into two groups: thirty African-American males and thirty European-American males. The subjects in the two groups were matched according to age and years of education. No males were taking hormonal supplements. A starting date for men in study 2 was determined at the subject's convenience.

The fragrances and the placebo were provided by International Flavors and Fragrances (Union Beach, New Jersey) in individual glass spray bottles for the females in study 1 and individual glass non-spray bottles for the males in study 2. All fragrances tested are currently on the market. The five glass bottles in each study were labeled "A" through "E." In the placebo condition, a solvent without fragrance was used. The placebo resembled the fragrances in appearance, and was labeled "F."

Three instruments were used to collect the data: 1) an initial questionnaire, 2) the POMS questionnaire, and 3) rating sheets for recording preference measures for the odor of the fragrances and placebo. Females also completed checklists for rating physiological symptoms throughout the study. Demographic, dietary, and medical information about each subject was obtained with the initial questionnaire. Mood ratings were obtained twice daily using the POMS questionnaire. The POMS consists of feelings which are rated on a scale from 0 (not at all) to 4 (extremely). Most of the 65 feelings can be divided into six independent factors: tension-anxiety, depression-dejection, anger-hostility, vigor-activity, fatigue-inertia, and confusion-bewilderment. From these factors, a Total Mood Disturbance score (TMD) can be derived by summing the tension, depression, anger, fatigue, and confusion factors and subtracting the vigor factor.

Preference ratings were obtained at the same time as the POMS for both the fragrances and the placebo. Subjects indicated their preference on a scale from 0 (dislike a lot) to 10 (like a lot). Female subjects completed the survey of physiological menopausal symptoms at three separate intervals throughout the study. The survey asked the severity of ten symptoms: hot flashes, night sweats, insomnia, headaches, fatigue, nervousness/irritability, joint pains, depression, dizziness, and other. Subjects rated these symptoms on a four-point scale labeled: none, slight, moderate, and marked. All subjects were given a brief questionnaire at the end of the study asking how they felt the fragrances used during the experiment had smelled on them, if they had noticed changes in how fragrances smelled on them over the years, if they had changed the amount of fragrance they had used over the years, and any other general comments.

Both studies lasted 12 days and were divided into three conditions: baseline, fragrance, and placebo. The results of the first two days were used as baseline information, and no fragrances were used. The following ten days consisted of five days of the fragrance condition where subjects would use the fragrances labeled "A" through "E," and five days of the placebo condition where subjects would use the placebo labeled "F." Thus, each subject served as his/her own control. For each group of fourteen women in study 1 and each race in study 2, half participated in the fragrance condition first and half participated in the placebo condition first.

Subjects filled out a POMS survey twice daily (midmorning and late afternoon) for all twelve days of the study. In addition, subjects completed the preference scales rating both the fragrances and the placebo. For both the fragrance and the placebo conditions, subjects applied the fragrances themselves and immediately filled out a POMS survey. For the fragrance condition, subjects were instructed to use as many or as few of the fragrances as they wished, according to their preference. For the placebo condition, subjects used only the placebo. At the beginning of the study and at the end of each condition, female subjects in Study 1 also completed the survey of physiological menopausal symptoms.

Data Analysis

In study 1, an analysis of variance was performed to determine if there were any significant main effects or interactions between the four groups of females and the conditions (baseline, fragrance, and placebo) for each POMS factor and the TMD. Both group and condition (baseline, fragrance, placebo) were found to have a significant effect upon mood (i.e. each POMS factor and the TMD). An interaction between hormonal group and condition was found for anger, vigor, and fatigue factors, and the TMD indicating that the scores for these factors were dependent upon both hormonal group and condition. Additionally, a special analysis was done which compared the average of the scores for groups 2 and 3 (hormonal group) with the scores of group 1 (menstruating group) and of group 4 (nonmenstruating group) for each POMS factor and the TMD. For the comparison between the hormonal group and the menstruating group, the hormonal group did significantly better than the menstruating group for all POMS factors and the TMD. For the comparison between the hormonal group and the nonmenstruating group, the hormonal group again had significantly better scores for all POMS factors, except anger, and the TMD. A Chi-square indicated that there were no significant changes in physiological menopausal symptoms from the baseline to the fragrance or placebo conditions.

In study 2, an analysis of variance was performed to determine if there were any significant main effects or interactions between the 2 races and the conditions (baseline, fragrance, placebo) for each POMS factor and the TMD. An effect of condition on mood scores was found for all POMS factors and the TMD.

An effect of race on mood sores was found for the tension, depression, and fatigue factors. European-American subjects had significantly higher scores for tension and fatigue than African-American subjects; however, African-American subjects had significantly higher scores for depression. An interaction between race and condition was found for depression, vigor, and confusion factors indicating that, for those three factors, mood scores were dependent upon both race and condition.

An analysis of variance was also performed comparing the data from both study 1 (females) and study 2 (males). The results revealed a significant effect of gender for all POMS factors and the TMD, but no interaction between gender and condition. Males had worse mood scores than females for tension, depression, anger, fatigue, and confusion factors, as well as for the TMD; males also had better scores than females for the vigor factor. An additional analysis was performed that compared mood scores for males with scores for women of taking hormone replacement therapy (either estrogen alone or estrogen with progesterone). Women taking hormone replacement therapy had significantly better scores than men for all POMS factors and the TMD. However, when scores for males were compared with scores for women who were no longer menstruating and taking no hormone replacement therapy, there were no significant differences in the scores for tension, fatigue, and confusion factors, as well as the TMD. Middle-aged females taking no hormonal supplements, however, still had significantly better scores for depression, anger, and vigor factors than males at mid-life.

Main Findings

The main finding of these studies is that the use of fragrance can improve mood in both women and men at mid-life. Feelings of tension, depression, and confusion were significantly alleviated by pleasant odors in female subjects. The finding that women taking hormonal supplements have better mood scores than those who do not is consistent with previous studies that have found gonadal steroids including estrogens elevate mood (Greenblatt, et al., 1979). Use of colognes significantly improved mood in men at mid-life (total mood scores as well as all individual factors of tension, depression, anger, vigor, fatigue, and confusion). The improvement of mood in men after exposure to pleasant odors was comparable for African-Americans and European-Americans for tension, anger, and fatigue factors, as well as for the TMD.

Several factors may account for the finding that pleasant odors can improve mood at mid-life. First, the area of the brain that processes olfactory information overlaps anatomically with areas that process information on emotions (Schiffman, 1983; Schiffman, et al., 1979). Second, Vliet and Davis (1991) suggested that fluctuations in mood at the female climacteric may be due to the effect of declines in hormone levels on central nervous system neurotransmitters including serotonin, norepinephrine, dopamine, and endorphin receptor systems. Thus, pleasant odor sensations may improve mood around female menopause, in part, by releasing a broad range of neurotransmitters in the limbic system. Third, learning and cognitive factors may play a role in the enhancement of mood if the odors have previously been associated with pleasant memories (Erlichman & Bastone, 1992; Lorig, 1992).

The finding here that females at mid-life (aged 45-60 years) have better total mood scores than males at mid-life (aged 40-55 years) is consistent with the findings of Lehr (1966) who studied the problems and conflicts in men and women that occur during the 5th and 6th decades. For women, the period from 40-49 years was stressful but the period after 50 was remarkably free of stress and conflict. Males, however, had a significant increase in stress beginning at 50 years of age. Perez-Milian (1991) reported that males in their 50's had increased feelings of failure and intensified their efforts to reach their goals. However, these intensifying efforts ceased by age 60.

While the moods for women were better as a group than moods for the males, the differences were largely accounted for by women taking hormonal supplements. Women taking either estrogen alone or estrogen with progesterone had better scores for tension, depression, anger, fatigue, and confusion factors, as well as for the TMD than males. When post-menopausal women who were not taking estrogen supplements were compared to males, there were fewer differences, and male and female subjects had comparable scores for tension, fatigue, and confusion factors. Post-menopausal women not taking hormonal supplements, however, still had better scores than males on depression, anger, and vigor factors, as well as the TMD. These data suggest that women have slightly better moods at mid-life since no males in the study were taking hormonal supplements.

Conclusion

Daily use of fragrance can significantly improve mood in women and men at mid-life. This improvement in mood is due to the pleasant odor.

References

Barbo, D.M. (1987). The physiology of the menopause. Med Clinics North Amer, 71(1), 11-22.

Calil, H.M., Zwicker, A.P. & Klepacz, S. (1990). The effects of lithium carbonate on healthy volunteers: mood stabilization. Biol Psychiatr, 27, 711-22.

Cockerill, I.M., Nevill, A.M. & Lyons, N. (1991). Modelling mood states in athletic performance. J Sports Sci, 9, 205-12.

Cole, J.O., Pope, H.G., Jr., LaBrie, R. & Ionescu-Pioggia, M. (1978). Assessing the subjective effect of stimulants in casual users. Clin Pharmacol Therap, 24, 243-52.

de Lignieres, B. & Vincens, M. (1982). Differential effects of exogenous oestradiol and progesterone on mood in post-menopausal women: individual dose/effect relationship. Maturitas, 4, 67-72.

Der, D.F. & Lewington, P. (1990). Rational self-directed hypnotherapy: a treatment for panic attacks. Am J Clin Hypnosis, 32, 160-7.

Ehrlichman, H. & Bastone, L. (1992). The use of odour in the study of emotion. In: van Toller, S. & Dodd, G.H. (Eds.) Fragrance. The psychology and biology of perfume. 143-59. London: Elsevier Applied Science.

Ehrlichman, H. & Halpern, J.N. (1988). Affect and memory: effects of pleasant and unpleasant odors on retrieval of happy and unhappy memories. J Personality Soc Psychol, 38, 213-28.

Featherstone, M. & Hepworth, M. (1985). The history of the male menopause 1848-1936. Maturitas, 7, 249-57.

File, S.E., Bond, A.J. & Lister, R.G. (1982). Interaction between effects of caffeine and lorazepam in performance tests and self-ratings. J Clin Psychopharmacol, 2, 102-6.

Flint, M. (1975). The menopause: reward or punishment? Psychosomatics, 16, 161-3.

Greenblatt, R.B., Nezhat, C., Roesel, R.A., & Natrajan, P.K. (1979). Update on the male and female climacteric. J Am Geriat Soc, 27, 481-90.

Holte, A. & Mikkelsen, A. (1991). Psychosocial determinants of climacteric complaints. Maturitas, 13, 205-15.

Horswill, C.A., Hickner, R.C., Scott, J.R., Costill, D.L. & Gould, D. (1990). Weight loss, dietary carbohydrate modifications, and high intensity, physical performance. Med. Science Sports Exer, 22, 470-6.

Hunter, M. (1992). The south-east England longitudinal study of the climacteric and postmenopause. Maturitas, 14, 117-26.

Jacobelli, A. (1985). The male climacteric. Endocrinologic profile and therapeutic prospectives. Clinica Terapeutica, 112, 155-61.

Janczewski, A., Bablok, L. & Czaplicki, M. (1967). Premature male climacteric. Polish Endocrinol, 18(1-2), 33-9.

Kalmar, H., Brandstatter, N. & Resinger, E. (1992). Psychological disorders of menopause: on the topic of multifactorial origin. Wiener Medizin Wochenschr, 142, 104-7.

Kantor, H.F., Milton, L.J. & Ernst, M.L. (1978). Comparative psychologic effects of estrogen administration on institutional and noninstitutional elderly women. J Amer Geriatr Soc, 26, 9-16.

Kraemer, R.R., Dzewaltowski, D.A., Blair, M.S., Rinehardt, K.F. & Castracane, V.D. (1990). Mood alteration from treadmill running and its relationship to beta-endorphin, corticotropin, and growth hormone. J Sprts Med Physical Fit, 30, 241-6.

Lehr, U. (1966). Problems and conflicts of middle age. Probleme und Ergebnisse der Psychologie, 16, 41-5.

Lieberman, H.R., Corkin, S., Spring, B.J., Growdon, J.H. & Wurtman, R.J. (1982/83). Mood, performance, and pain sensitivity: changes induced by food constituents. J Psychiat Res, 17, 135-45.

Lorig, T.S. (1992). Cognitive and non-cognitive effects of odour exposure: electrophysiological and behavioral evidence. In: van Toller, S. & Dodd, G.H. (Eds.) Fragrance. The psychology and biology of perfume, 161-73. London: Elsevier Applied Science.

McNair, D.M. & Lorr, M. (1964). An analysis of mood in neurotics. J Abnorm Soc Psychol, 69, 620-7.

McNair, D.M., Lorr, M. & Droppleman, L.F. (Revised 1992) Manual: Profile of mood states. San Diego: Education and Industrial Testing Service: San Diego, CA, 1992.

Nankin, H.R. (1985). Fertility in aging men. Maturitas, 7, 259-65.

Padwick, M.L., Endacott, J. & Whitehead, M.I. (1985). Efficacy, acceptability, and metabolic effects of transdermal estradiol in the management of post-menopausal women. Am J Obstet Gynecol, 52, 1085-91.

Parlee, M.B. (1978). Psychological aspects of the climacteric in women. Psychiatr Opinion, 15, 36-40.

Perez-Milian, R. (1991). Males' sexual and life transitions, awareness, and climacteric. MA thesis. University of West Florida.

Polit, D.F. & LaRocco, S.A. (1980). Social and psychological correlates of menopausal symptoms. Psychosom Med, 42, 335-45.

Rausch, J.L., Nichinson, B., Lamke, C. & Matloff, J. (1990). Influence of negative affect on smoking cessation treatment outcome: a pilot study. Br J Addict, 85, 929-33.

Schiffman, S.S. (1983). Taste and smell in disease. New Engl J Med, 308, 1275-9;1887-943.

Schiffman, S.S., Orlandi, M., & Erickson, R.P. (1979). Changes in taste and smell with age: biological aspects. In: Ordy, J.M. & Brizzee, K. (Eds.) Sensory systems and communication in the elderly, 247-68. New York: Raven Press.

Schindler, B.A. (1987). The psychiatric disorders of mid-life. Med Clinics North Amer, 7(1), 71-85.

Schmidt, P.J. & Rubinow, D.R. (1991). Menopause-related affective disorders: a justification for further study. Am J Psychiatry, 148, 844-52.

Soules, M.R. & Bremner, W.J. (1982). The menopause and climacteric: Endocrinologic basis and associated symptomatology. J Am Geriatr Soc, 30, 547-61.

Szarvas, F. (1992). Male climacteric from a practical point of view. Wiener Medizinische Wochenschrift, 142, 100-3.

Vartapetov, B.A. & Demchenko, A.N. (1965). The climacteric in men. Kiev USSR: Zdorov'ya, 244.

Vliet, E.L. & Davis, V.L.H. (1991). New perspectives on the relationship of hormone changes to affective disorders in the perimenopause. NAACOGs Clinical Issues in Perinatal & Women's Health Nursing, 2, 453-71.

Wiklund, I., Holst, J., Karlberg, J., Mattson, L.A., Samsioe, G., Sandin, K., Uvebrant, M. & von Schoultz, B. (1992). A new methodological approach to the evaluation of quality of life in postmenopausal women. Maturitas, 14, 211-24.

Williams, T.J., Krahenbuhl, G.S. & Morgan, D.W. (1991). Mood state and running economy in moderately trained male runners. Med Science Sports Exer, 23, 727-31.

The Author
Susan Schiffman

Dr. Susan Schiffman is Professor of Medical Psychology and Director of the Weight Loss Unit in the Department of Psychiatry at Duke University Medical Center in Durham, North Carolina. She is also the Chairperson of the Olfactory Research Fund's Scientific Advisory Committee. Dr. Schiffman is an internationally recognized authority on taste and smell and their role in nutrition and human behavior. Her research spans the range from clinical to molecular investigations of the senses of taste and smell. Dr. Schiffman studies how odors affect our daily lives. She also studies how taste and smell affect feeding behavior and nutrition in the elderly and the obese. Dr. Schiffman has published extensively in numerous scientific journals.

Cultural and Historical Perspectives

physiology • olfactory c

d social behavior • mood • cultural and historical perspectives • applications • fundamentals of odor perc

VI

and psy

entals of

perception • fragrance and psych

Experimental studies sponsored by the Olfactory Research Fund have produced findings that will have an impact beyond the confines of the research laboratory. To make the most of these and other empirical findings, it is useful to place them in a cultural and historical perspective. The Fund supported the efforts of David Howes, Anthony Synnott and Constance Clausen in the writing of "Essence", a book on the history, sociology and anthropology of smell, an overview of which is presented here.

The same investigators undertook an examination of the ethnographic data available for the cultures of Africa, Oceania, and South America. They found evidence for two distinct cultural uses of odors: classificatory and dynamic. The former uses odor to classify people and places into a coherent philosophical arrangement, while the dynamic use of odor is about ritual and transition, for example purifying, healing, and creating group identity.

These anthropological investigations show that the cultural use of scent is closely tied to its symbolic content. Taking a cue from this work, Anthony Synnott conducted a smell survey that used open-ended questions to probe for cultural knowledge about the sense of smell and its meanings.

ESSENCE: The History, Sociology and Anthropology of Odour

David Howes

Anthony Synnott

Constance Classen

David Howes
Anthony Synnott
Department of Sociology and Anthropology
Concordia University, Montreal, Canada

Constance Classen
Center for the Study of World Religions
Harvard University, Cambridge, Massachusetts

Most of the work on smell which has been undertaken to date has been of a physical or scientific nature. Smell, however, is not simply a biological phenomenon, it is a cultural phenomenon, and as such has a social history as well as a natural history. Our book, **Essence**, will offer the first comprehensive exploration of the cultural role of odours in different periods of Western history down to the present, and in a wide range of non-Western societies.

In the modern West, we tend to think of smells in purely aesthetic terms - they are either pleasant or unpleasant. In many cultures however, odours have provided and still provide a basic means of defining and interacting with the world. This is particularly the case in so far as odours are closely associated with personal and group identity. The study of the history, anthropology and sociology of odour is, in a very real sense, an investigation into the "essence" of human culture itself.

In what follows, we present an overview of the contents of **Essence**. Part I, "In Search of Lost Scents," concerns the history of smell; Part II, "Explorations in Olfactory Difference" delves into the anthropology of odour; while Part III, "Odour, Power, and Society" is about the sociology of smell.

Part I. In Search of Lost Scents

Chapter 1.
The Aromas of Antiquity

Paul Faure writes in <u>Parfums et aromates de l'antiquite</u> that our senses of smell and taste are so under-developed in the modern West that we can no more appreciate the importance of odour in the ancient world than the blind can describe a colourful scene. This chapter tries to recover the olfactory landscapes of the ancient world by exploring the different roles played by aromatics during this period.

Fragrance was central to the cultures of antiquity. Perfumes and incense were extensively employed for practical, esthetic and religious purposes. Practical purposes included the use of fragrance as a purifier and fumigant. The esthetic value accorded to fragrance was so great that there was scarcely an element of domestic life - from food to clothes to pets - which was left unscented. Most important of all for the ancients was the religious role of fragrance: sacred scents were believed to attract and unite humans and gods, and thus integrate the cosmos in a harmonious olfactory whole.

Chapter 2.
Following the Scent:
From the Middle Ages to Modernity

After the fall of the Roman Empire, the use of perfumes fell into decline. With the Crusades, however, the peoples of the West were once again brought into contact with Eastern peoples, and with the spices and perfumes which had so entranced the Greeks and Romans. This chapter examines the spiritual and curative powers attributed to scent during the Middle Ages. It explores such topics as "the odour of sanctity," and the role played by smell during periods of plague. It then traces the rise of the art of perfumery beginning in the sixteenth century. Perfumes were an essential part of the high life of Renaissance and Enlightenment Europe. This was so much the case that, for example, court etiquette in seventeenth-century Versailles demanded that a different scent be worn each day of the week.

This emphasis on olfactory aesthetics was accompanied by a decline in the belief in the medicinal and spiritual value of scent. By the nineteenth century, fragrance had moved out of the realms of religion and science and into those of sentiment and sensuality. This move is brought out in the works of many writers of this period, such as Baudelaire, and later, Proust, who used olfactory symbolism in his novels to create an evocative atmosphere. The final section of this chapter examines the attitudes of nineteenth century philosophers and scientists - from Nietzsche to Freud - towards odour and explores their influence on the olfactory norms of the modern West.

Part II. Explorations in Olfactory Difference

Chapter 3.
Universes of Odour

This chapter explores how smell is used in non-Western cultures to structure and classify different aspects of the world, from time and space to gender and selfhood. Examples are drawn from the "osmologies" – or olfactory classification systems – of a wide array of cultures, including the Desana Indians of the Colombian Amazon, the Ongee of the Andaman Islands, the Temiar of Malaysia, the Trobriand Islanders of Papua New Guinea, the Dassanetch of Ethiopia, and the Chinese.

The chapter opens with an account of the "calendar of scents" used by the Ongee to reckon time, and moves on to examine the olfactory construction of space among various tribes of the Colombian Amazon. The role of smell in the development of a sense of selfhood among the Temiar is considered next, followed by a survey of the ways in which the idiom of smell is used to express social divisions (gender, age, occupation, etc.) in cultures from around the world. The chapter concludes with a discussion of the classificatory and aesthetic uses of smell in the courtly life and medical traditions of China. It is shown that smells were worked into an elaborate system of correspondences by ancient Chinese scholars, a system that included the seasons, the cardinal points, different organs of the body, government ministries, and so on.

Chapter 4.
The Rites of Smell

Aromatics are employed in different cultures for a variety of purposes, including seduction, healing, hunting, and communication with the spirits. The forms which these olfactory customs take are highly diverse. In the Trobriand Islands, for example, a perfume made of coconut oil blended with mint is considered an essential aid to attracting the attentions of a potential lover,

while among the Dassanetch of Ethiopia the smell of cow manure is considered aphrodisia-cal. With regard to healing, the Warao of Venezuela, who may be said to enjoy the most sophisticated system of aromatherapy in the world, enlist powerful herbal scents to combat the evil odours of disease. Among various peoples of the interior of Papua New Guinea known for their hunting abilities, a hunter sleeps with an bundle of herbs tucked under his pillow. The aroma of these herbs is supposed to inspire dreams of the chase. The next day he has only to act out his scent-inspired dream to enjoy a successful hunt. The Bororo of the Brazilian Amazon seek the favour of the beneficial spirits of their religion by offering them the sweet scents of corn, white mud and duck feathers. Evil spirits, in turn, are said to feed on putrid odours, the elimination of which is thus of constant concern to the Bororo.

These ritual uses of scent animate the olfactory structures discussed in the previous chapter, providing a means of both integrating and distinguishing the different spheres of life.

Part III: Odour, Power and Society

Chapter 5.
The Politics of Smell

This chapter takes as its point of departure an observation made by George Orwell. According to Orwell,

> the real secret of class distinctions in the West is summed up in four frightful words ... *The lower classes smell* ... No feeling of like or dislike is quite so fundamental as a *physical* feeling.

Orwell goes on to state that race hatred, religious hatred, differences of education, of temperament, and so on, can all be gotten over, "but physical repulsion cannot," whence the persistence of class distinctions.

Orwell's point is well taken, but while physical feeling may seem fundamental, it is important to underline that its basis in the case

he described remains social rather than physical, as class divisions are given in society, not in nature. This point introduces us to the politics of smell. Olfactory preferences and repulsions tend to be very deep seated. Manipulating such preferences, therefore, is a common and effective means of generating and maintaining social hierarchies, perpetuating a given political ordering. This explains why odours are often invoked not only to create and enforce class boundaries, but also to police ethnic, gender, and religious boundaries. By examining different instances of the operation of such olfactory codes in the modern West, the interrelations of odour, power and society are revealed.

Chapter 6.
Olfactory Management

The regulation of smells is a subject of intense concern in contemporary Western society.

This regulation involves the suppression of odours that are perceived as unpleasant, and the elaboration and use of pleasant scents. Olfactory management takes place on numerous levels: the body, the home, the workplace, the marketplace and the environment. At the level of the body, for instance, deodorants suppress unwanted odours while perfumes and colognes allow for the creation of an ideal olfactory image. At the level of the workplace, the concern is how to develop an attractive olfactory atmosphere that will stimulate and refresh workers, as opposed to the stale air that is usually found in the enclosed modern office building. In the marketplace, businesses are increasingly concerned not only with new ways of marketing perfumes, such as home fragrance products and aromatherapy, but with the addition of appealing fragrances to a variety of products - from running shoes to paint. In the urban environment, "olfactory pollution" has become a problem as foul odours produced by industrial complexes invade residential neighbourhoods. The question arises of what constitutes an acceptable olfactory environment, and that is precisely the question this final chapter seeks to answer.

The Authors

Anthony Synnott

After brief stints in the navy and the Jesuits, Dr. Synnott studied in Rhodesia (as it then was), at the London School of Economics, and received his Ph.D. from London University. He is now researching the body and the senses and teaching sociology at Concordia University in Montreal. His book, The Body Social: Symbolism, Self and Society, was published by Routledge in 1993.

In addition to collaborating as researchers on the Olfactory Research Fund sponsored project, "The Power of Smell in Historical, Sociological and Anthropological Perspective," Constance Classen, and David Howes have written a book, Aroma: The Cultural History of Smell, published by Routledge in 1994.

David Howes

Dr. David Howes holds a Ph.D. in Anthropology from the University of Montreal, and two degrees in Law from McGill University. He has carried out field research in Papua New Guinea and Northwestern Argentina, and is presently teaching anthropology at Concordia University. He is the editor of The Varieties of Sensory Experience: A Sourcebook in the Anthropology of the Senses (University of Toronto Press, 1991), and the author of numerous articles.

Constance Classen

Dr. Constance Classen has a Ph.D. in Religious Studies from McGill University. She has undertaken extensive research on concepts of the body and the senses in the indigenous cultures of South America and in Western history. Her publications include Inca Cosmology and the Human Body (University of Utah Press, 1993) and Worlds of Sense: Exploring the Senses in History and Across Cultures (Routledge, 1993).

The Anthropology of Odour

David Howes

Anthony Synnott

Constance Classen

David Howes
Anthony Synnott
Department of Sociology and Anthropology
Concordia University, Montreal, Canada

Constance Classen
Center for the Study of World Religions
Harvard University, Cambridge, Massachusetts

Cultures differ in the meaning and importance they attach to the different senses. This fact is especially apparent in the case of smell. In The Hidden Dimension, anthropologist E.T. Hall states with respect to our own culture that:

> *In the use of the olfactory apparatus Americans are culturally underdeveloped. The extensive use of deodorants and the suppression of odour in public places results in a land of olfactory blandness and sameness that would be difficult to duplicate anywhere else in the world. This blandness makes for undifferentiated spaces and deprives us of richness and variety in our life. It also obscures memories, because smell evokes much deeper memories than either vision or sound.*

Are there cultures which have a more developed sense of smell and in which life is therefore richer? What are the cultural uses of smell?

These questions inspired us to examine the ethnographic record on the cultures of Africa, Oceania and South America in order "to smell what we could smell." We discovered that the power of smell has indeed been put to many different and creative uses in the cultures of those lands.

In the first and second parts of this paper, we report on what those uses are, though we do so only in summary fashion due to limitations of space. For the full flavour and aroma of these uses, the reader is referred to *"SMELL AND CULTURE: An Annotated Bibliography of Sources on the Indigenous Cultures of Africa, Oceania, and Latin America,"* which was submitted as part of the Final Report on the "Anthropology of Odour" Research Project. In the third part of this paper, we draw out some of the implications of our findings for the fragrance industry. That industry, we believe, has a fundamental role to play in reawakening Americans to the wisdom, power and pleasures of the senses - above all, smell. Odour, as a medium of communication and channel of personal expression, has unlimited potential for growth.

It is useful, for purposes of analysis, to distinguish between three kinds of odour. An odour can be either *natural* (for example, body odour), *manufactured* (for example, perfume), or *symbolic* (for example, the belief that each race has a distinct odour – a scientifically untenable proposition). It is also useful to distinguish between the *classificatory* and *dynamic* uses of odour. The term "classificatory" refers to the use of smell as a basis for ordering the world – that is, for distinguishing between different classes of people, animals and things. The term "dynamic" refers to the use of odour in ritual and everyday contexts, often with a view to changing the world, or restoring it to its proper state.

I. Olfactory Classification

The six most basic uses of odour for classificatory purposes may be summarized as follows:

1) Classifying people, animals and plants by their natural odour.

2) Classifying people, animals and plants by the symbolic odors attributed to them. For example, it is commonly supposed that different races each have a different smell, and even that "the 'other' race stinks" - but there is no empirical evidence to support this belief.

3) Classifying groups within a society; for example, men and women, children and adults, by natural and symbolic odors.

4) Classifying space by reference to the environmental odour of different territories.

5) Classifying the cosmos through odour. For example, assigning contrasting symbolic odors to sun and moon (as among the Batek Negrito of Malaysia), or odorizing fundamental cosmic and social principles such as "structure" and "change" (as among the Bororo of Brazil).

6) Establishing a value system based on olfactory symbolism. For example, characterizing certain odors as good or bad, and assigning them to different beings or states, in order to signify the latter's moral goodness or badness.

As an example of a typical non-Western olfactory classification system, consider the system employed by the Suya of Brazil. The Suya classify animals by odour, rather than, say, morphology or habitat. The same terms that are used to classify animals are used to classify people, and to a lesser degree, plants. While animals are permanently classified in a given category, human beings have different odours according to sex, stage in the life-cycle, and transition through certain ambiguous states, such as initiation or illness.

As Anthony Seeger states in <u>Nature and Society in Central Brazil</u>: "The categorization of the [natural and social] world in terms of odour provides an important system for the interpretation of Suya actions and attitudes." Thus,

> *The most powerful and important animals in the Suya cosmology are all strong smelling, while the less important ones are pungent or bland. Human beings are not all equally social. Men are socialized through initiation and lose their strong-smelling odour. Women, on the other hand, by their very sexuality are strong smelling. Old people are neither as fully social as adult men nor as sexually marked as young women, and old males and females are both pungent.*

The level of olfactory consciousness among the Suya is, evidently, much higher than among ourselves. The principal reason for this is that for the Suya smells have meaning; they do not simply provoke reactions of pleasure or disgust, the way they do for us. To put this another way, the Suya think in smell, whereas we only react to smells, because our culture does not provide us with a framework in terms of which to think of odours as symbolic vehicles. Colors can symbolize concepts for us, as in the case of the traffic light system, where red means "stop," green means "go," and so on. Sounds also have meaning for us, for example, the soundtrack of a movie tells us what emotions we should be feeling as the action unfolds. But odours are not coded by our culture (or more likely, the code has been forgotten), and this deprives us of any model in terms of which to organize our olfactory experience. Hence, our response to smells can only be measured in terms of relative pleasure. Of course there is nothing stopping our society from re-developing an olfactory code, but this would require a more integrated and totalizing production and marketing strategy on the part of the fragrance industry than exists at present. This matter will be taken up again after the next section.

II. Dynamics of Smell

The twelve most salient uses of odour for dynamic purposes may be summarized as follows:

1) Establishing group identity through some odour, whether natural, manufactured, symbolic, or some combination of these. For example, East African pastoralists, such as the Dassanetch, smear themselves with cattle products to give themselves a bovine scent. This odour of cattle differentiates them as a group from neighbouring fishermen.

2) Communicating messages through odours. For example the use of different sorts of incense to establish channels of communication with different spirits, each spirit being associated with a specific scent.

3) Employing odours as a means of attraction, whether of members of the opposite sex, game animals, or spirits.

4) Employing odours as a means of repulsion, whether of enemies, animals or evil spirits.

5) Employing odours to enhance one's chances for success at a particular endeavour, such as in playing games of chance.

6) Employing odours in order to cleanse and purify, both in ritual and practical contexts, either as an alternative to or in conjunction with the use of water.

7) Employing odours to heal, both directly through the administration of curative smells, and indirectly by creating a pleasant olfactory environment for the patient.

8) Employing odours in rituals of transition, such as weddings and funerals.

9) Employing odours as a means of establishing exchange relations with other persons and groups. For example, giving and receiving products with different odours in rituals of exchange, best exemplified in the Desana practice of exchanging ants of different odours.

10) Employing odour to direct experience. For example, using odoriferous substances to inspire particular kinds of dreams, to guide a person through a hallucinogenic trance, or to suppress memories of the deceased at a funeral.

11) Attributing the power of olfaction to plants (as among the Wamira of New Guinea) and inanimate objects (as among the Kwoma, also of New Guinea), or attributing an extremely discerning nose to the gods (as among the Batek Negrito of Malaysia), and explaining misfortune in terms of said plants, objects or gods taking offence at the mixing of odours which results from people engaging in forbidden activities.

12) Employing olfactory metaphors to express abstract concepts and values, such as the idea of an "odour soul" among the Temiar.

III. Implications for the Fragrance Industry

The preceding discussion of the uses of odour in the indigenous cultures of Africa, Oceania, and South America can be analyzed from various angles with a view to extracting the marketing strategies that should guide the fragrance industry into the 21st century.

Briefly, the four most salient strategies which the fragrance industry could, with profit, dedicate itself to implementing are as follows:

A. Diversification of Perfume and Perfume Use;

B. Development of Aromatherapy as an Alternative Medicine;

C. Development of Olfactory Awareness and Symbolism;

D. Integration of Odours with Other Sensory Stimuli.

In what follows we shall elaborate on what each one of these strategies or initiatives could involve.

A. Diversification of Perfume and Perfume Use

As the ethnographic literature surveyed in Smell and Culture suggests, many niches are occupied by odours in the lives of the peoples of Africa, Oceania and Latin America. Each of these niches could in theory be developed into a new "market niche" in the context of North American society. A list of such niches, some of which have already no doubt been tried, would include:

i) The creation of a range of scents that suggest different natural environments, such as the seashore, the rainforest, the smell of the earth after rain.

ii) The creation of more sophisticated scents for the home environment (instead of simply pine, lilac, etc.).

iii) Different aromas for different rooms of the house, for the workplace, school, stores, etc.

iv) Different aromas for different age and cultural groups.

v) Different aromas for different members of the family.

vi) Different aromas for different parts of the body.

vii) Different aromas for different moods, viz. pensive, joyful, sensuous.

viii) Different aromas for different seasons, times of day, kinds of weather, viz. autumn, dawn, rainy.

ix) Different aromas for different endeavours, viz. job hunting, romance, sports.

x) Different aromas for different events, viz. birthdays, weddings, funerals.

xi) The creation of mix-and-match fragrances – that is, aroma kits that include guidelines for the combination of scents which the consumer may use to create his or her own personal "atmosphere."

B. Development of Aromatherapy as an Alternative Medicine

In increasing numbers, North Americans are turning to medical systems other than Western biomedicine, in search of well-being. The rejection of biomedicine and the embrace of East Asian medicine (acupuncture, acupressure, moxibustion), homeopathy, and, above all, aromatherapy, arises out of a growing concern over the many side-effects of surgical and chemical interventions, as well as a growing "cult of the natural."

Many aromatherapy clinics have sprung up across the face of North America, but North American practitioners of aromatherapy tend to be grossly ignorant of the olfactory traditions of those societies, such as the Warao of Venezuela, where aromatherapy is really a science. Among the Warao, the inside of the body is conceived of as a kind of gas pressure chamber, where all sorts of olfactory reactions take place. Diagnosis is by smell rather than x-rays or the chemical analysis of blood samples, such as one finds in biomedicine, and treatment is by the application of scents.

The olfactory wisdom of the Warao, and other groups like them, needs to be studied in depth, and the principles of their system of olfactory healing should be publicized. Just as the pharmaceutical industry has profited greatly from studying the pharmacopoeias of South American Indian shamans, so may the fragrance industry be able to profit from popularizing the aromatherapies of South American Indian herbalists. The first priority, however, is to sponsor research teams to collect some of this wisdom. David Howes and Constance Classen report on their research among Andean *curanderos* (traditional healers) in "Aromatherapy in the Andes," which appeared the 6/1993 issue of the Dragoco Report.

C. Development of Olfactory Awareness and Symbolism

The challenge of seeking to right the "underdevelopment of smell" in North America is a daunting one, but not impossible. What would seem to be required is the following:

i) Encouraging olfactory education of children at the primary school level.

ii) Promoting interest in perfumery as an art form.

iii) Encouraging people to develop classificatory systems based on odour. For example, coordinating fragrances with astrological signs.

Regarding the first suggestion – namely, making olfactory education a part of the curriculum (in the same way that, say, physical education is recognized as an essential component of every child's development), a useful model is provided by the experience of the French. Since January 1991, primary school students throughout France have had their curriculum enriched by a range of courses dedicated to educating their sense of taste. While the principal aim of this educational program has been to inculcate in the students a taste for French haute cuisine (as opposed to American fast-food), the program has also had the effect of refining their sense of smell. The students have developed into "aromands" as well as "gourmands." The Fragrance Foundation has already experimented with the design of an Aroma Kit. Designing an expanded version of this kit for use in schools could have the highly desirable outcome of training the next generation of noses to be more discriminating in the attention they pay to the odours that surround them.

The third suggestion – namely, encouraging people to begin coordinating fragrances with other phenomena, such as astrological signs – is not as far-fetched as it may sound. Many cultures have elaborated Tables of Correspondences, the Chinese Table being perhaps the most famous. In the Chinese system, each of the elements (wood, metal, earth, fire, water) out of which the cosmos and the human body is composed has a corresponding color, flavor, musical note, and odour, as well as direction, season, and so on. To try and construct such a system of correspondences can prove both intellectually and sensually gratifying. There is much to commend such schemes, from our perspective, for not only do they give smells meaning, but they also integrate the sense of smell with all the other senses, which brings us to the last marketing strategy that we wish to propose:

D. Integration of Odours with Other Sensory Stimuli

North American society and culture has traditionally been dominated by the visual faculty: "Oh say can you see ..." goes the U.S. national anthem. Increasingly, however, there are signs of a "rejection of visualism" taking shape. "Visualism" is a frame of mind which, like racism and sexism, is gradually being eroded. The overthrow of the hegemony of the visual will result in the liberation of the other senses, and the emergence of new sorts of selves. What we particularly look forward to are the following possible developments:

i) The creation of multisensory works of art and entertainment, viz. combining music with fragrance.

ii) The creation of perfumes inspired by a specific work of art: a painting, piece of music or poem.

iii) The inclusion of fragrance as an integral part of home decoration. For example, employing a specific fragrance to match the decor of each room.

iv) The use of fragrance in previously non-odorized products, viz. jewelry, giftwrap.

These developments are possible because they have already been tried out by diverse non-Western cultures, which proves that they are latent in the human condition - just waiting to be expressed in late twentieth-century America.

References

Classen, C., Howes, D. (1993). Aromatherapy in the Andes. Totowa, NJ: Dragoco, Inc.

Hall, E.T. (1969). The Hidden Dimension. New York: Doubleday - 1977. Beyond Culture. New York: Anchor Books.

Seeger, A. (1981). Nature and Society in Central Brazil: The Suya Indians of Mato Grosso. Cambridge, Mass.: Harvard University Press.

The Authors

Anthony Synnott

After brief stints in the navy and the Jesuits, Dr. Synnott studied in Rhodesia (as it then was), at the London School of Economics, and received his Ph.D. from London University. He is now researching the body and the senses and teaching sociology at Concordia University in Montreal. His book, The Body Social: Symbolism, Self and Society, was published by Routledge in 1993.

In addition to collaborating as researchers on the Olfactory Research Fund sponsored project, "The Power of Smell in Historical, Sociological and Anthropological Perspective," Constance Classen, and David Howes have written a book, Aroma: The Cultural History of Smell, published by Routledge in 1994.

David Howes

Dr. David Howes holds a Ph.D. in Anthropology from the University of Montreal, and two degrees in Law from McGill University. He has carried out field research in Papua New Guinea and Northwestern Argentina, and is presently teaching anthropology at Concordia University. He is the editor of The Varieties of Sensory Experience: A Sourcebook in the Anthropology of the Senses (University of Toronto Press, 1991), and the author of numerous articles.

Constance Classen

Dr. Constance Classen has a Ph.D. in Religious Studies from McGill University. She has undertaken extensive research on concepts of the body and the senses in the indigenous cultures of South America and in Western history. Her publications include Inca Cosmology and the Human Body (University of Utah Press, 1993) and Worlds of Sense: Exploring the Senses in History and Across Cultures (Routledge, 1993).

Roses, Coffee and Lovers: The Meanings of Smell

Anthony Synnott
Department of Sociology and Anthropology
Concordia University, Montreal, Canada,

What is smell? The concise Oxford Dictionary explains that it is the: "Nasal sense by which odours are perceived." And an odour is defined as a "Pleasant or unpleasant smell." Yet odours are far more than pleasant or unpleasant; and this nasal sense means far more than simply odour perception, as we discovered when we surveyed 270 students and faculty at Concordia University in Montreal about what the sense of smell means to them. The questions were open-ended and included such topics as the ranking of smell in the sensory hierarchy, our favourite (and also most disliked odours), how smell affects our personal relations, the relations between odour and memories, and the patterns and processes of fragrance use[1].

The Hierarchy of the Senses[2].

The first two questions of the Concordia Smell Survey asked the respondents which one of the five senses is MOST precious to you? Which one is LEAST precious to you? And why?

Sight is overwhelmingly the most valued (80%), followed remotely by touch (13%) and hearing (7%) - and only three people, one percent of the sample, rated smell as their most precious sense. Why is smell so precious to them? One person explained: "It is the sense that brings back the most vivid memories"; nonetheless, his choice is strange for he is a musician and might therefore be expected to value hearing the most. Another explained this point:

> *Because my sense of smell has actively worked in creating and stimulating my memories. Lilacs, appleblossoms, that cold, crisp smell of an early spring night. Evergreen*

trees, the glorious aromas of food cooking in the kitchen - all are kept alive by my sense of smell (45-year-old female "homemaker").

Conversely, the least valued sense - the one people would be most willing to lose if they had to lose one - is smell (55%), followed by taste (35%), touch (5%), hearing (4%) and sight (2%); almost an exact reversal of the most valued hierarchy.

It is noteworthy that each one of the senses is most valued by some (except taste), and most disvalued by others. Evidently we all do live very different sensory lives; but the hegemony of sight and the low value of smell are the only senses on which there is a majority consensus.

Why does smell rank so low? These responses are representative of this majority consensus:

Aroma doesn't interest or please me too much. If I can get by without smelling during a cold, I could get by indefinitely.

Smell...is precious of course, but being a smoker, I can hardly smell anything anyhow.

I have temporarily lost my sense of smell due to a chemical accident. I was able to get along without it, although I was very relieved when it returned.

Many people have no sense of smell, or a poor sense of smell, sometimes due to smoking or allergies; and most people have lost their smell, at least temporarily, due to colds and flus; and since they survived without it, they believe "I could get by indefinitely." Smell (and taste, which is physiologically linked to smell) are

probably the only senses which people regularly lose: we do not go blind temporarily every couple of years, nor do we lost our sense of touch. Deafness can occur but it may be temporary, due to wax build-ups or after air travel. In general, however, smell is "seen" (a word which "reflects" the hegemony of sight) as the least useful sense and also the one people are most accustomed to losing. One individual, who is anosmic, has learned to smell in colour, an interesting excercise in synaesthesia. Since the questionnaire was anonymous, this process could not be investigated further, but the attitude to smell is worthy of further research:

I lost my sense of smell in an accident and aside from many foods tasting bad, I learned to compensate and mentally visualize certain tastes by colour images. I have relearned smell again although often if something smells bad I can't tell unless someone else tells me it's bad.

There has been some considerable discussion in the literature that women and men live different sensory lives: Women are tactile (a proximity sense): warm, nurturing and so on, while men are ruled by sight (a distance sense): cold, straight, with blind spots, unfeeling, rigid. The debate obviously is highly political, and will not be presented in detail. An analysis of the Concordia sensory hierarchy by gender, however, shows no major quantitative differences in the ranking of the most valued senses; but in the ranking of the least valued senses the women preferred to lose taste most (44%) followed by smell (40%) and touch (8%), whereas the men preferred to lose smell first (58%), followed by taste (31%) and touch (3%).

Favourite Smells

Despite the low ranking of smell on the sensory totem pole, many people do enjoy a rich appreciation of smell. Asked "What are your favourite smells?" some of the

respondents were quite lyrical, and they were obviously highly sensitive to smells. Some respondents emphasized "people smells" first:

My favourite smells are "people smells" - The unique smell each person has. My favourite smell sensation is when I smell a familiar smell which brings me right back to a person from my past.

Men (fresh out of a shower) or first thing in the morning, babies, freshly rained on pavement, wood fire, cats' noses, ocean...camomile. They are calming and soothing. They remind me of moments in time that I would otherwise have forgotten. They make me happy.

Body perspiration - is individual, natural, an aphrodisiac. The smell of babies. The crisp smell of spring-time. These are both fresh and clean smells that both represent the beginning of a new life and season for us.

Body odours (except someone who hasn't washed in a long time), sea air, hot earth. Probably because smells have a huge array of associations, memories – they seem to contain much more information than the other senses.

The smell of new-born babies, baby powder, oranges (all fruit), the sea (fishy smell), aloe vera lotion, freshly fallen snow (and the smell of a cold, crisp night), freshly cut grass, the beach, coconut (one of my favourites), cinnamon, basil, parsley, the smell of my home and the smell of a hot summer day. All of these smells (except for baby powder and suntan lotion) are smells of nature – and are therefore natural, simple and real. These smells give me a sense of euphoria!

These smells are favourite because they entice memories from the past, especially of people, because they are aphrodisiacal, because they contain "information" and because they

cause "euphoria." Other smells are similar in their consequences.

One individual is evidently a nose:

The smells of my lover's hair, neck, beard, eyes, etc., etc., etc., Each little part smells wonderfully different. Some heartwarming, some soothing, some exciting. I still remember vividly and miss the smell of some of the men in my life: spice, sunshine, grass, dampness... .
The warm smell of my baby and later of my little boy when he came in from outside, all sweaty and sun-drenched.
I think there is no true <u>emotional</u> bonding without touching and smelling, burying one's nose into a loved one. (That even goes for furry pets.)
I never realized how much I made love with my nose.

The theme of emotional bonding and loving was noted by others (see below), and many people commented on how favourite odours affect their moods. Apart from "people" smells, many respondents discussed a wide range of "natural" smells of roses, flowers, the ocean, tomatoes, eucalyptus... . The word most frequently used in assessing their appreciation of those aromas was "fresh."

Roses, flowers, leaves, fresh air, food lasagna, perfume, peppermint tea leaves, oranges, cinnamon, spices. Roses smell so fresh, they relax me, force me to inhale deeply.

Freshly mowed lawn in the summer; fresh flowers in June and July (roses, carnations). It gives me a feeling of relaxation and relief from the hibernation I have to endure during the long winter season.

Flowers, chocolate, candle burning. They simply relax me. It changes my mood, makes me feel pleasant.

The ocean smell in the morning, the smell of a summer rainy day, in fall the smell of leaves, food cooking. They give me pleasure and a sense of security that I don't find in anything or anywhere else. Perfumes make me sneeze.

Tomatoes fresh from the garden, the fragrance of petunias caught on the softest summer breeze, cat's fur, that wonderful smell of burning leaves in the autumn. Because they bring back wonderful memories of being outside as a child.

Eucalyptus. It has a cleansing quality. As you inhale, the smell penetrates your lungs and it feels as if it is healing.

Fresh baking bread and fresh air...the fresh air makes me "feel clean, more alive," it kind of lights my spirits.

Fresh grass, good food, rain on a hot summer's day. I feel alive, "reborn" when I smell rain, fresh grass – I think perhaps it has to do with newness – I like new beginnings.

Wet autumn leaves. Campfires. My grandmother used to build huge bonfires for the entire family. Now I pit fire my own pottery in a bonfire, so the scent is still special for me.

Many people referred to such (relatively) artificial odours as cooking and fragrances. But most people cited a mix of natural and artificial odours; and, even more interestingly, many people could not separate their favourite aromas from the memories associated with them. Smell and memory are tangled up together deep in the unconscious:

Home-made baked beans, home-made bread. We used to eat these when I lived with my parents. We took them to the woods to camp out on Sundays. My Dad warmed up the beans over an open fire. The smell of beans and wood was everywhere.

Coffee. I have no idea why I like the smell of coffee, because I don't even drink the stuff.

Men's cologne, good food cooking, grass, dogs; especially puppy breath, spring (wet grass, mud, flowers, etc.). All of these things remind me of either a safe haven of childhood or something I took pleasure in. Cooking reminds me of my mother. Not just any kind of food, but the kind of food she cooked. Like potatoes and fried onions. I loved being outside in the summer and I always had dogs that I loved a lot so I love any kind of smell of theirs; wet fur; even their shit doesn't offend me.

Talcum powder – smells like my grandmother. Gasoline – reminds me of all the places I can go and have been in my car i.e., freedom. Water – the smell of the ocean or a lake, or the ground after it has rained, remind me of my home in B.C.

Roses, various fruits (apples, peaches) freshly cut grass, steak, my boyfriend's natural odour. Roses, fruits and grass because they have a fresh but mildly sweet fragrance. Steak because it whets my appetite. My boyfriend's odour because it triggers emotions of love in me e.g., if I find a tee shirt he has worn that has his odour clinging to it, I will immediately feel awashed in love.

"Awashed in love" – some people are obviously much more sensitive to smells than are others. Many of us are missing a range of sensory possibilities that others enjoy. Perhaps the most striking aspect of these responses is the degree to which odours affect moods and feelings. Time after time the respondents mentioned "feeling alive," "invigorated," "serene," "pleasant," "relaxed," "relieved," "calming and soothing," "happy" as well as "awashed."

Smells are mood-altering and entirely legal, not fattening and often free. These analyses imply that we should pay far more attention to odours, and in so doing gain a little more control over our sentient lives, and live a little more pleasurably.

Horrible Smells

Roses, coffee and lovers are not the entire story. The list of smells that people find obnoxious is long. Some are natural: body odour and sweat, vomit, morning breath, dirty feet, farting, human and animal waste products, skunks and rotten fish. Perhaps the reader is feeling nauseated already.

Smelly men on the bus in summer. It makes me feel as if they don't care about themselves or the people around them.

Odours of mildew and fungi and waste (garbage), simply because they get me nauseous.

BAD body odour!! It makes me sick to my stomach. I associate it with uncleanliness.

I dislike wet, musty smells - either from feet or wet clothing. It reminds me of cold snow and being uncomfortable - all things which I don't like.

Filthy odours because they are in my perception of smell (as awful) representing the negatives or should nots of the world.

Some odours are artificial: cigarette and cigar smoke, exhaust fumes, various cooking odours, pollution, pulp and paper mills,

Some responses were distinctly odd. One man's favourite smells were:

The odours of the Montreal Forum and the Olympic Stadium. Although some people would describe these smells as verging on foul, I love them because I associate them with my two favourite sports: hockey and baseball.

garbage and decay, sulfur, ammonia...

The smell of hospitals. Raw meat. The hospital smell reminds me of death and mortality. Raw meat smells like murder and killing.

Exhaust, perfumes – because they are poisonous.

Pollution/city odours. Because not only are they foul, they are unhealthy.

Anything chemical and anything irritating like cigarette smoke or heavy perfume.

These quotations are particularly interesting as equating bad odours and bad health: a union of chemistry and morality. These negative odours are also underline physically powerful:

These odours are annoying to my nose and either make me sneeze or feel sick to my stomach.

They are nauseating and result in severe headaches.

They make me sick.

Body odour and bad breath. They usually make me sick to my stomach.

Indeed these physical reactions to negative effluvia are precisely as strong as the positive emotional reactions to pleasant aromas. Smell is clearly not simply odour perception, but an intrinsic component of our emotional lives, and indeed our physical worlds. One person concluded that bad odours are not only unpleasant but also painful and "can't be good for you":

> Burnt pie in the oven, things growing/rotting in the fridge or elsewhere, pig farms and chicken coops, rabbit urine, vomit. They are extremely unpleasant. They hurt your nose (really!). If your nose doesn't like it: you know it can't be good for you.

Oddly enough, many of the most disliked odours were designed precisely to be liked, to be aromatic.

> Totally loathe bathroom and car deodorant.

Personal Realities

Friedrich Nietzsche was one of the few philosophers to discuss smell, but what he did say brought personal relations down to basics:

> What separates two people most profoundly is a different sense and degree of cleanliness. What avails all decency and mutual usefulness and good will toward each other – in the end the fact remains: "They can't stand each other's smell!" (Nietzsche, 1886/1966).

We asked if they had experienced this. About one quarter of the sample said "No;" but one enthusiast replied briefly "I love body odour." Many people complained generally about body odour on public transit systems; but some were much more specific:

> Yes! I don't really know how to explain it, but some people's smell (no matter how clean they are) is

> Perfumes. Must be allergic. Instant headache and nausea. All smell like industrial cleaners.

> Perfume. It desensitizes one's senses and lacks 'diversity' of smells. (Perfumes cover up other smells in the area.)

There seems to be a fair consensus on which odours are most disliked; though what is most striking are the physical reactions to bad odours – nausea, headaches, migraine, feeling sick; also the emerging equation of bad odours with "bad" people (the careless smelly man on the bus), "bad" situations (hospitals) and "bad" health (poisonous, unhealthy air pollution). Bad odours symbolize moral badness of one type or another, and also cause physical badness i.e., illness. Odours are powerful metaphors of reality.

> not attractive. You have no desire to snuggle up to something that makes one of my senses unhappy.

> Yes, my grandmother has very strong body odour and for some reason doesn't use deodorant. I love her dearly but her odour is so strong sometimes I literally feel sick to my stomach.

> Some teacher in high school never changed his shirt – I never got extra help I needed. He made me cringe when he walked by.

> My closest friend's breath is so awful that at times I feel the need to distance myself from her.

> I have worked with some co-workers who do not use deodorant – the heavy smell of body odour can be a bit overpowering.

Can't really be attracted to a woman who smokes cigarettes. The smell penetrates clothing. Like licking an ashtray.

One respondent seemed to think we were casting aspersions: "Never! because all my friends are clean." No doubt they are; but who said anything about friends? Smell is still a "sensitive" topic. Others protested against the (excessive?) deodorization and sanitization of the human body:

Cleanliness is a bourgeois concept. I think it is mad how we go about "covering up" our own natural bodily odours. I get to know others

(somehow) by how they naturally smell.

The United States or North America in general is so caught up in hygiene that the natural smells of the body disgust them, they are offended by their own bodies.

That may be; people are offended by body odours they define as foul. In a recent study of public transportation users in the United States, 46% objected to bad-smelling people; and, for the sake of completeness, 50% were annoyed by people wearing heavy perfume, loud talking (49%), getting too close (48%) and crying babies (46%). What intrigues us is the 50% or so who are not annoyed by these things!

Memories

Helen Keller, who knew about smells, was lyrical about their power, especially for activating memories:

Smell is a potent wizard that transports us across a thousand miles and all the years we have lived. The odour of fruits wafts me to my Southern home, to my childish frolics in the peach orchard. Other odours, instantaneous and fleeting, cause my heart to dilate joyously or contract with remembered grief. (Keller, 1908).

More concise was Rudyard Kipling who wrote that "Smells are surer than sounds or sights to make your heart-strings crack." So we asked, "Do odours trigger your memories?" Many people did not answer, but whether due to time constraints or other reasons, we do not know. Some, however, were ecstatic:

Yes, yes, ex-lovers re-encountered. Their sight, sound and touch are wonderful, but nothing compares to their smell.

Yes. Yes - psychologically triggered memories from childhood, comforting,

arise from smell of mother's scarf.

Yes they do! Just this weekend I was putting a coconut cream on and I had a vivid memory of when I was in Mexico last year putting the same cream on.

Yes. Certain colognes remind me of ex-boyfriends and they put me in a daze remembering my past.

I find that I have a strong sense of smell and that odours trigger my memories predominantly more than a picture or sound might. I have become very attached to the body odour of my boyfriend. Not odour from perspiring but his own distinct smell. I can trigger memories of my boyfriend and I doing things even if I just sniff his clothes or jacket.

Definitely. Every time I smell Magie Noire by Lancome, I think of my mother. She wears it all the time!

And how! Perfume goes straight to your heart, by-passing the censure of the mind, leaving you breathless.

Yes. Brut 33 will always remind me of my first love, and Stetson reminds me of when he changed colognes and dumped me (but I don't think the cologne had anything to do with it).

Gasoline – When I was young my Dad was a mechanic (when he lived with us). Perfume – I wore with my first boyfriend especially mixed with his, 'Old Spice,' still makes me feel in love.

Wonderful that odours still make some people "feel in love." What else could? And what price for such a feeling? But although most of the memories cited are happy ones, some are sad.

Yes, my mother's clothing (she died recently). Her things are my only physical link to her. To smell them brings back the presence of her.

Yes, odours really do trigger memories. I was on the bus one day and a man got on the bus and was sitting in front of me. All of a sudden I began to think of my Grandfather and realized that I was thinking of him because I could

smell the cologne he always used to wear. It made me sad because he had just died recently.

When my father passed away two years ago, we put a certain kind of flower in front of his picture. That same kind of smell reminds me of sadness, the helplessness, worst of all my mother's crying.

Memory is selective, and so are odours. One person observed this:

Odours trigger good memories not bad ones. The smell of the ocean triggers many good memories of times I spent on the beach. Windex doesn't evoke the times I spent in detention at school washing desks.

This point is well taken, as Proust (1982) most clearly explained.

The lesson from these aromatic inhalations, surely, is that we should be more alert to the emotional implications of our olfactory universe: we should, following Nietzsche, smell more, and even consider fragrancing our environments for our deeper emotional satisfaction.

Earliest Memories

We also enquired about people's earliest memories of smell. Many people recalled their mother, and sometimes the positive emotions attaching to their mothers. These two quotations are representative:

I guess it would have to be my mother, she always smelled like baby powder, that really clean, soft smell.

The earliest smell I can remember is my mother's perfume. She still wears the same one but I always remember that smell when she would lean over and kiss me and hug me. There would always be that same smell and it always

meant the same thing – comfort.

Some people's first memories are of food:

I can remember the smell of coffee and frying bacon. We used to go camping when I was very young and we would cook outside – they were great times.

Baking. My mother baked a lot when I was young. The smell is comforting.

These odours were usually, but not always, positive and happy. Carrots may be problematic:

I do believe it was mashed carrots. I

hate the taste and smell of carrots. I can remember when I was 2 or 3, my mother tried to make me eat them. I 'slapped' the spoon and hid my face. That's when I got carrots all over my face. The smell of carrots reminds me of that.

Some people objected vigorously to their own odors:

I would have to say my diaper days. I knew I wanted that diaper off.

While others remember their first memories in negative terms, and other people's odours:

The smell of my Dad smoking Export A in the car and the smell of his

lighter. (We used to travel a lot between Montreal and Toronto.) [This same woman says the odour she dislikes most is cigarette smoke.]

Some recalled the natural country odours of their youth: pines, farms, grass, and manure.

Manure, because both my parents worked and I lived with my grandparents on the farm and spent much of the time with the horses.

First smell memories ranged widely from mothers to milk, cotton sheets to coal fires, incense to the sea, grass to cigarettes. Most seem to be happy memories, but some do seem to be more comforting than others.

Fragrances

Personal fragrance is a $2.5 billion industry in the United States alone (Standard and Poor, 1992), but the total aroma industry, including the fragrance components of shampoos, detergents, soaps, air fresheners, polishes, the food industry, even car interiors, is probably closer to $25 billion. I don't think anyone knows. Aromas are certainly economic.

Aromas are highly personal. We asked the respondents which fragrances they wore, why and how often. The responses were extremely varied: tastes and attitudes differed, and so did the practice.

To start with use: some people wear fragrances every day, mostly women; some never wear them, particularly men. Women are twice as likely to use fragrances as men always, or almost always; and men are almost twice as likely never, or rarely, to use fragrances. Two thirds of women use fragrances always or almost always, compared to one-third of men. Most women wear them, most men don't.

What fragrances did these men and women favour? There is no clear consensus.

Some liked many fragrances, some liked none at all, and some couldn't smell anything anyway. The two most popular fragrances for women were: Anais Anais and Lauren. Men are less likely to wear fragrances, and if they do use them, they do so less often. The two top fragrances for men were Polo and Drakkar.

There were some controversies, however. What some people loved, others loathed. One person's favourite scent is patchouli: "Patchouli is a calming, warm and woodsy smell. It smells real and earthy." Another described it laconically as the most disliked odour: "Patchouli = mold, old." Polo is the favourite cologne of some men, and violently disliked by others. "Polo: it was worn to death." "Old Spice, Brut, Polo - they're too strong, gag reflex."

Why do you wear fragrances? and when? Answers varied, naturally. Some women wear it for themselves, some for others, some for both, and others never. Fragrance is highly political, therefore; and these responses give some idea of the range, but also the functions, of these fragrances:

To feel pretty.

It makes me feel good, clean, pretty, feminine.

Boyfriends have liked it in the past.

Often I wear my favourite scent when I want to be sure to leave a memory like at a job interview or a special date.

After showers – I feel cleaner, fresher. Evenings out – makes my boyfriend wild plus leaving him for the evening makes my scent stay with him all night.

Everyday, because I like to smell nice, and have people comment that I'm wearing a nice smelling perfume. My boyfriend will come up to me and say "hmmmm, you smell so wonderful."

I believe I wear perfume to feel somewhat more attractive. Socialization has led me to believe I'll feel prettier – and it works!!

To feel pretty, to attract, to make him wild, to stay behind and because "it works!!" These are some of the functions for women. Men are usually less eloquent on the topic, and less subtle too.

To go out to bars in order to pick up.

Every morning so I can feel fresh and start the day right.

Whenever I go out for a more formal evening, or to mask the fact I have not showered if I am late in preparing for work.

But some people reject fragrances and perfumes totally. When do they wear them?

Never. I'm allergic to most of them and really couldn't be bothered finding one that would agree with me. That would entail my having to spend time next to products that make me miserable!!

I never wear them. As far as I'm concerned they all smell awful. What's wrong with soap and water?

It was a liberating moment for me last year to stop wearing any deodorant under my armpits. To break the barrier of convention. I wear oils on my neck so my 'b.o.' is not really that well masked. My smell (I think) is not that offensive, but it does have a definite, distinct scent.

Some say "it works!!"; others object to it on political grounds, and still others are allergic to many artificial fragrances.

Indeed today attitudes to odours in general, and fragrances in particular, seem to be polarizing. On the one hand we are paying increasing attention to the richness and range that odour plays in our personal and social lives: that fragrances do "work" for many people: they make them feel clean or fresh or pretty or feminine, as the case may be, and they do attract others; fragrancing of the environment, as pioneered by the Japanese, does affect moods, and also production; odours are a powerful, subliminal marketing tool: coffee odours wafted on to the street draw in customers, fragranced products often sell better than unfragranced, the smell of leather sells new, as well as old cars, and so on. Indeed Annette Green of the Olfactory Research Fund believes that the future of the fragrance industry lies in the research relating the different fragrances to moods and behaviors: "Fragrances which reduce stress, increase alertness, improve social relationships, encourage relaxation and even sleep will dominate" (Green, 1991).

On the other hand there is also increasing opposition not only to the negative effluvia of air-pollution generated by industries, as well as to carcinogenic cigarette smoke, but also to perfumes. Some have protested the use of scented magazine inserts, and the indiscriminate spraying of customers in stores and shopping malls. "Perfume Pollutes" is the slogan of one New Jersey health group. (The opposition retorts: "I have a right to choose how to make myself smell") (Toronto Globe and Mail, March 17, 1993).

Activists in the United States have launched protest movements advocating the banning of perfumes and fragrances from public meetings and public places. Organizations involved include the Atlanta-based *Human Ecology* *League* and the *National Centre for Environmental Strategies* in New Jersey. The logistics of these strategies may prove difficult to implement - but demands for a fragrance-free environment and for no-scent areas of restaurants, together with smoking and non-smoking sections, will no doubt escalate in the future (Toronto Globe and Mail, March 17, 1993). Evidently there is no pleasing everybody. Fragrance is in the nose of the beholder.

Conclusion

Smells do many things: cause headaches and stomach sickness, and also happy moods; they arouse memories and take us back in time; they may be therapeutic, or not. But surely above all they are not well understood or appreciated. Indeed, completing this questionnaire itself evoked a re-evaluation of this sense, to the surprise of many of the respondents:

Ironically after completing this questionnaire I realize how important smell is to one's life.

It is ironic that smell is the sense I said I wished to lose if I had to choose, when obviously it is very important. I am going to quit smoking so I can smell things again.

After this questionnaire I'm going to try and be more aware of my sense of smell.

After answering these questions, I guess smell is more important than I thought.

Respondents were asked to add any anecdotes about smell that they could think of. Some have been already incorporated into the discussion, as seemed appropriate; but others are so individual that they did not "fit."

I try to discourage my children from eating sweets or chocolate, but having a sweet tooth of my own, I *generally indulge myself once they have gone to bed. One evening one of my sons called me for whatever reason just as I was enjoying my chocolate bar. I went to his room leaned over his bed and asked him what the problem was. He immediately asked if I had been eating chocolate. I said no, but he kept sniffing harder and told me he was sure I had been eating a chocolate bar because he could smell it.*

My two-year-old son, Jon, has an incredible interest in smells. He loves smelling spices and anything I use for cooking. The other day he pulled out his father's aftershave, (which he never uses, so no association there). I gave him my perfume bottles and he decisively rejected them in favour of the 'manly' aftershave.

The only anecdote I have is that whenever I'm in the car and I smell something strange – I'll shut off my radio to smell better – makes no sense!!!

The last word goes to the fellow who noticed the relationship between humans and animals in a way that surely Darwin would have been proud of:

I've always been kind of fond of sniffing pets as a display of affection. How odd that they feel (smell?) the same way about me.

Not odd really: we are animals, and we do sniff out the world, perhaps more than Hegel and Freud believed; we are more animalistic than we know, and more than we appreciate. It is especially important for liking and disliking: "I never knew how much I made love with my nose." The nose is more important than we know.

Notes

[1] I would like to thank first all the students and faculty at Concordia who participated in our survey; also David Howes and Constance Classen, my partners in this research in the sensorium; my colleagues Joseph Smucker and Pearl Crichton, for their support; Carole Robertson again for her superb typing services; Nicholette Starkie, who did so much work on the distribution and analysis of the surveys; and finally to Helene Tobin, who helped sniff things out.

[2] This, and other issues, are discussed in more detail in David Howes (Ed.) The Varieties of Sensory Experience (Toronto: University of Toronto Press, 1991); Anthony Synnott, The Body Social (New York: Routledge, 1993); Constance Classen, Worlds of Sense: Exploring the Senses in History and Across Cultures. (New York: Routledge, 1993); Annick Le Guerer, Scent (Translated by Richard Miller. New York: Turtle Bay Books, 1992).

References

Annette Green. (1991). "New Sensory Era Beckons America's Fragrance Industry." Dragoco Report, 2, 39-42.

Helen Keller, "Sense and Sensibility." Century Magazine, 75, February 1908, 566-77, 773-83.

Friedrich Nietzsche, Beyond Good and Evil. (Translated by Walter Kaufmann. New York: Vintage Books, 1966).

Marcel Proust, Swann's Way. (Translated by C.K. Scott Moncrieff. New York: Modern Library, 1982).

The Author
Anthony Synnott

After brief stints in the navy and the Jesuits, Dr. Synnott studied in Rhodesia (as it then was), at the London School of Economics, and received his Ph.D. from London University. He is now researching the body and the senses and teaching sociology at Concordia University in Montreal. His book, The Body Social: Symbolism, Self and Society, was published by Routledge in 1993.

In addition to collaborating as researchers on the Olfactory Research Fund sponsored project, "The Power of Smell in Historical, Sociological and Anthropological Perspective," Constance Classen and David Howes have written a book, Aroma: The Cultural History of Smell, to be published by Routledge in the Fall of 1994.

Applications

VII

Many of the rigorous scientific studies sponsored by the Olfactory Research Fund document effects of fragrance on basic psychophysical processes and mental states. One challenge that arises is to put these findings to work in a practical, applied setting. <u>Mary Kliauga, Kathy Hubert and Terri Cenci</u> have taken up the challenge in their report about the effects of ambient fragrance on proofreading efficiency.

Consumer Panel Study on The Effect of Peppermint and Lavender Fragrances on Proofreading Efficiency

Mary Kliauga   *Kathy Hubert*   *Terri Cenci*

Mary Kliauga, Director of Bureau of Chemistry and Environmental Studies
Kathy Hubert, Director of Beauty Laboratory
Terri Cenci, Beauty Laboratory
Good Housekeeping Institute

Introduction

The Good Housekeeping Institute Bureau of Chemistry and Environmental Studies and Beauty Laboratory, working under a grant from the Olfactory Research Fund, conducted consumer panel studies to ascertain the effect of peppermint and lavender fragrances on proofreading efficiency. Current Aroma-Chology research suggests that peppermint fragrance may have an alerting effect on performance while the lavender fragrance may have a relaxing effect. Proofreading requires alertness and concentration, while at the same time it is perceived to be a tedious task. During our study, sixty-seven (67) volunteers recruited from the Hearst Corporation proofread pages of text containing misspelled words that they were to identify. Various peppermint, lavender and control (no fragrance) sessions were conducted. The efficiency of proofreading was compared between fragrance and no fragrance. We scented the office conference room using a diffuser, the type used by consumers at home. The overall results showed that participants performed significantly better when fragrance was diffused throughout the room compared with no fragrance at all. Lavender fragrance had a statistically significant effect within 95% confidence level (Phase I study). The effects of fragrance vary between persons and depend strongly on an individual's preference. Although the men's panel size was small, we found that men performed better with peppermint fragrance, while women performed better with lavender fragrance.

Procedure

Our study consisted of two parts: Phase I and II. The Phase II study was conducted in an effort to randomize fragrances. We recruited volunteers employed at various jobs at the Hearst Corporation. The participants were men and women of all ages from 20 to over 60 years old. They were secretaries, editors, office managers, chemists, engineers, home economists, and technicians.

The study was blind, in that the participants were not told that the rooms were scented or not scented (control). They were told they would participate in a proofreading study. The volunteers were asked to proofread ten to fifteen pages of short stories containing misspelled words deliberately built into the text. The participants were asked to identify any misspelling by circling or underlining the word. The stories were selected from past issues of Good Housekeeping magazine and Reader's Digest. Approximately 10-15 volunteers proofread for 30 minutes in a room that is typical of an office environment.

We scented the room using a commercially available fragrance diffuser, the type used by a consumer at home (BioRegene Aromatic Diffuser Model III, Leydet, Fair Oaks, CA). This diffuser has low and high settings. We used the low setting so that the fragrance was not overpowering. The weight of fragrance used was monitored by weighing the diffuser before and after the proofreading sessions. The fragrance oil was diffused throughout the room during 30 minutes of proofreading.

We used peppermint oil (NF Natural VS, CAS No. 8006-90-4) and lavender oil (40/42, NF CAS No. 8000-28-0) purchased from Polarome Manufacturing Company, New York, NY.

We compared the proofreading scores under peppermint and lavender fragrances with no fragrance. We counted the total spelling errors identified (circled or underlined) and subtracted the words incorrectly identified. Proofreading scores were calculated as:

FIGURE 1

Proofreading score for Phases I and II with Lavender fragrance

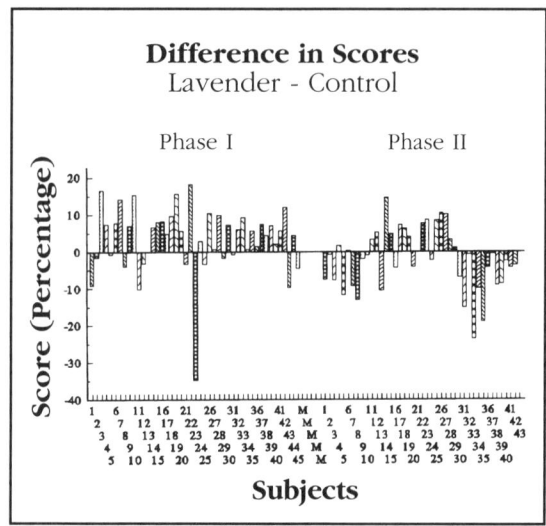

FIGURE 2

Proofreading score for Phases I and II with Peppermint fragrance

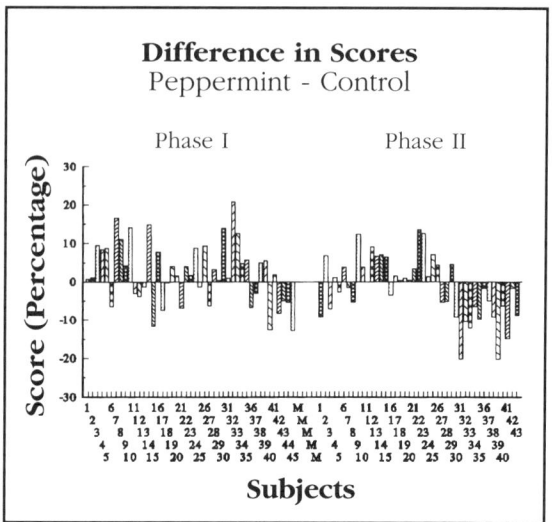

% Errors found (Score) = $\dfrac{(A - B) \times 100}{C}$

where   A = Total number of errors identified

B = Number identified incorrectly

C = True number of misspelled words in the text

To compare individual performance scores for fragrance compared to the control (no fragrance) we subtracted the control score from the fragrance score, and plotted the differences in a bar graph (Figs. 1 and 2). Each bar represents one individual. The upper half of the graph shows those who scored higher than control and the lower half of the graph shows those who scored worse than control.

The panelists proofread in a room scented with peppermint, control and lavender fragrance. The Phase I study results showed that the lavender fragrance improved the average proofreading scores significantly at a 95% confidence level (Fig. 1, Phase I). Although peppermint increased average scores of proofreading, the difference was not as great.

However, the overall effect was positive. Many participants improved scores as much as 15-20% (Fig. 1, Phase I). We can conclude from the Phase I study that the results of this initial effort are by no means conclusive, but are certainly suggestive and intriguing. One of the most intriguing aspects can be observed by noticing that some individuals are far more sensitive to fragrance than others. Some, in fact, appear to have a negative response to lavender fragrance (e.g., Fig. 1, Participant No. 23). Very clear and dramatic positive response to peppermint fragrance was exhibited by certain individuals (e.g., Fig. 2, participants No. 32 and 7).

The Phase II Study was randomized in that the order in which participants received control, lavender or peppermint was different for each group. A total of 43 volunteers participated. We divided panelists into three groups (A, B, and C). Group A and B volunteers had participated in the Phase I study. Group C people were mostly new panelists. The fragrance treatment and story texts were randomly distributed across groups as shown in Table 1.

FIGURE 3

Phase II scores for Group A, B, and C with Lavender

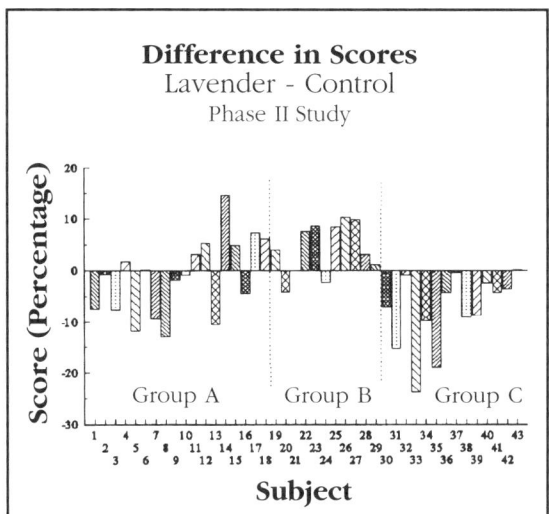

FIGURE 4

Phase II scores for Group A, B, and C with Peppermint

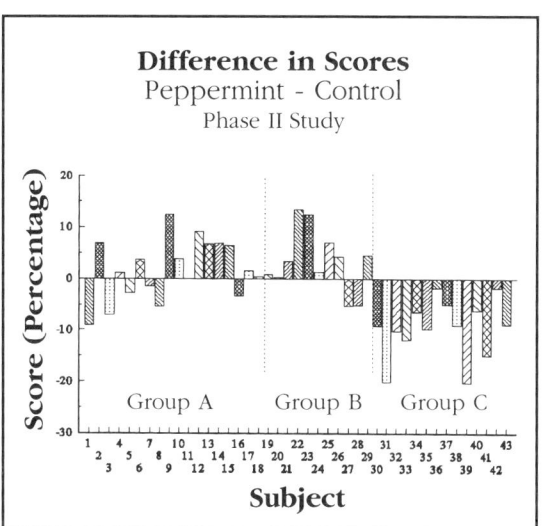

TABLE 1

Group A	Group B	Group C	Story
1. Peppermint	2. Lavender	3. Control	Story #1
4. Lavender	5. Control	6. Peppermint	Story #2
7. Control	8. Peppermint	9. Lavender	Story #3

Phase II results showed that the performance of Groups A and B (70% of panel) were consistent with the Phase I study. The fragrance improved proofreading. However, we found no significant difference in mean scores during Phase II. This is because, for some unknown reason, Group C people (subjects 30-43) performed worse with both peppermint and lavender than control (no fragrance). As seen in Figs. 3 and 4, every individual in Group C had a negative score with both fragrances.

The reason for the discrepant result with Group C is not known. Data were re-examined for errors, but none were found. Group C fragrance scores were normal, but control scores were much higher than other groups which gave the negative effect. We need to further examine the data. Taken by themselves, the results of Groups A and B, consisting of 30 people, confirm our earlier results of Phase I that fragrances do tend to enhance performance of proofreading tasks.

In order to see the overall effects of fragrance on proofreading study, we pooled the Phase I and Phase II results. We have plotted men's and women's scores separately. The results will be discussed in the conclusion.

We believe that representing the data as a comparison of individual's performance with different fragrances, is better suited to the type of experiment that we did than a simple quantitative comparison of means and/or medians. Because the number of subjects and the number of sessions was relatively small, the mean can be greatly influenced by a single individual who may have scored extremely low or high to a certain fragrance.

Conclusion

We found some exciting results.

- Our study showed that fragrances improved proofreading efficiency.

- It is apparent that the effect of lavender fragrance was overwhelmingly positive (Fig.1).

- The effect of fragrance depends upon individual's preference. Some individuals responded more strongly to fragrance than others. We observed that some even have a distinct negative response to lavender or a consistent large positive response. With this type of representation of the data, one can follow a given individual's performance.

- Men and women appeared to have different preference of fragrances. Even though the number of men panel was small (9 men compared with 58 women), our data suggested an interesting effect. The men performed better with peppermint while women performed better with lavender fragrance.

Acknowledgement

This research was supported by a grant from the Olfactory Research Fund, to whom we express our appreciation.

The Authors

Mary Kliauga

Mary Kliauga is the Director of the Bureau of Chemistry and Environmental Studies at the Good Housekeeping Institute. She has a BS degree in chemistry and MS degree in cosmetic science from Fairleigh Dickinson University. She is a member of the American Chemical Society, Society of Cosmetic Chemists, American Society for Testing And Materials, Association of Official Analytical Chemists.

Kathleen Hubert

Kathleen Hubert is the Director of the Beauty Laboratory at the Good Housekeeping Institute. She attained her Masters degree in 1987 and is a member of the Society of Cosmetic Chemists and the Dermal Clinical Evaluation Society, and The Fashion Group

Providencia Cenci

Providencia (Terri) Cenci is a licensed cosmetologist in the Beauty Laboratory at the Good Housekeeping Institute. She is a member of the Society of Cosmetic Chemists, The Fashion Group, and National Cosmetology Association. She is a communication major at Hunter College.

The Aroma-Chology Review...

*Focus on Fragrance and
Olfactory Developments and
Breakthroughs around the World*

THE AROMA-CHOLOGY* REVIEW... is the only source of information in lay language which reports on the results of current scientific including studies supported by the Olfactory Research Fund in the field of Aroma-Chology. The publication also tracks current international news in the medical, technical, physical and social sciences which may be applied to the study of the sense of smell and the psychological benefits of fragrance.

THE AROMA-CHOLOGY REVIEW... is published semi-annually and contains articles from a wide diversity of sources; olfactory scientists, researchers and authors of the medical profession from around the world.

THE AROMA-CHOLOGY REVIEW... is available by subscription. For further information, please contact the Olfactory Research Fund, 145 East 32nd Street, New York, NY 10016. Fax number: (212) 779-9058

*Aroma-Chology is a registered mark of the Olfactory Research Fund

NOTES

NOTES

NOTES